STUDY GUIDE

Mindy Miller-Kittrell, Ph.D.
North Carolina State University

Elizabeth A. Machunis-Masuoka, Ph.D.
University of Virginia

Microbiology

with Diseases by Body System

Third Edition

Robert W. Bauman

Exam I

Ch. 1
Ch. 3
Ch. 5
Ch. 6
Ch. 7
Ch. 8

Benjamin Cummings
Boston Columbus Indianapolis New York San Francisco Upper Saddle River
Amsterdam Cape Town Dubai London Madrid Milan Munich Paris Montreal Toronto
Delhi Mexico City Sao Paulo Sydney Hong Kong Seoul Singapore Taipei Tokyo

Executive Editor: Leslie Berriman
Development Director: Barbara Yien
Project Editor: Katie Seibel
Supplement Editor: Denise Wright, Southern Editorial and Production, Inc.
Senior Managing Editor: Deborah Cogan
Production Manager: Kathy Sleys
Senior Marketing Manager: Neena Bali
Project Management/Composition: Moganambigai Sundaramurthy/Integra
Main Text Cover Design: Riezebos Holzbaur Design Group
Cover Photo: Visuals Unlimited/Corbis
Supplement Cover Design: 17th Street

1 2 3 4 5 6 7 8 9—[BRR]—15 14 13 12 11

Benjamin Cummings
is an imprint of

www.pearsonhighered.com

ISBN-10: 0-321-71629-9
ISBN-13: 978-0-321-71629-3

Contents

STUDY GUIDE

Preface

As with other sciences, microbiology can sometimes be difficult to learn. There is a great deal of material to master, and the field demands strong problem-solving skills. This study guide has been designed to help you develop the skills you need to meet these challenges.

The study guide corresponds chapter by chapter with *Microbiology with Diseases by Body System*, Third Edition by Robert W. Bauman. Each study guide chapter begins with a detailed Chapter Summary of the corresponding textbook chapter and can be useful in organizing your notes. Use these summaries as an aid in reviewing each chapter's material before attempting to answer questions in the textbook or study guide. The "Key Themes" section serves two purposes: to explain why you are studying each particular chapter, and to provide you with broad concepts to focus on as you study. The multiple choice, fill in the blank, and matching questions are designed to test your mastery of key terms, concepts, and microbial processes. Short-answer questions provide you with the opportunity to summarize fundamental concepts. Critical thinking questions build upon these fundamentals by asking you to apply concepts to solving problem-oriented questions. Finally, concept building questions are designed to have you synthesize information from multiple chapters in a comprehensive approach to understanding complex scenarios.

Answers to all questions appear at the back of the study guide. Always try to answer the questions before looking at the answers. Don't skip questions because they appear hard—it is important that you think through difficult concepts to build your problem-solving skills. If your answer is wrong, compare your answer with the one given in the textbook, and consult the textbook to help you understand and correct your answer.

It is our hope that you will find microbiology an exciting course of study. By strengthening your problem-solving abilities, you will be able to meet the challenges of your studies and, hopefully, have fun along the way.

Mindy Miller-Kittrell
North Carolina State University

Elizabeth A. Machunis-Masuoka
University of Virginia

Using the MasteringMicrobiology™ Study Area

In addition to this Study Guide, you can also access a variety of student study and review materials in the Study Area of MasteringMicrobiology (www.masteringmicrobiology.com).

WHAT IS THE MASTERINGMICROBIOLOGY™ STUDY AREA?

The Study Area in MasteringMicrobiology contains a wealth of resources designed to help you succeed in your microbiology course. You can get ready for tests with its simple, three-step approach:

1. **Take a Pre-Test** and obtain a personalized Study Plan.
2. **Learn & Practice** with animations, activities, and MP3 Tutor Sessions.
3. **Test Yourself** with quizzes and a chapter post-test.

You will also have access to a variety of self-study tools, including:

- **Get Ready for Microbiology** includes an eText of *Get Ready for Microbiology* by Lori K. Garrett and Judy Meier Penn and diagnostic tests to find out what you know and what you need to review.
- **Concept Mapping activities** help you practice building maps to organize concepts in a meaningful, visual way. These concept maps also appear in the end-of-chapter sections in the textbook.
- **MP3 Tutor Sessions** allow you to download MP3s for specific chapters of the textbook and study on the go. You can listen to mini-lectures about the tough topics and take audio quizzes to check your understanding.
- **MicroFlix™** are 3-D movie-quality animations with self-paced tutorials and gradable quizzes that help you master the three toughest topics in microbiology: metabolism, DNA replication, and immunology. You can view the animations, complete the tutorial, print a study sheet, and take the quiz. You can also access BioFlix™ animations to help you review relevant concepts from general biology.
- **115 multi-step Microbiology Animations** explain and visually demonstrate core concepts, providing an additional chance for you to learn. They are accompanied by gradable quizzes.
- **Microbiology Videos** consist of short video clips of microorganisms in motion, bringing microbiology to life in a way that text and photos alone cannot.
- **Microbiology on the Web** consists of links to recommended websites to supplement your reading of each chapter in the text.
- **Flashcards** allow you to review the key terms in the text in flashcard format. Cards may be sorted by chapter, and you can choose to review key terms from as few or as many chapters as you would like.
- **Careers** consists of information about professional possibilities in the field of microbiology.
- **EText** is your link to a convenient, online version of the textbook.

Study Guide

1 A Brief History of Microbiology

CHAPTER SUMMARY

The Early Years of Microbiology (pp. 2–7)

The early years of microbiology brought the first observations of microbial life and the initial efforts to organize them into logical classifications.

What Does Life Really Look Like?

Antoni van Leeuwenhoek (1632–1723), a Dutch tailor, made the first simple microscope in order to examine the quality of cloth. The device was little more than a magnifying glass with screws for manipulating the specimen, but it allowed him to begin the first rigorous examination and documentation of the microbial world. He reported the existence of protozoa in 1674 and of bacteria in 1676.

By the end of the 19th century, Leewenhoek's "beasties" were called **microorganisms**. Today they are also known as **microbes**.

How Can Microbes Be Classified?

During the 18th century, Carolus Linnaeus (1707–1778), a Swedish botanist, developed a system for naming plants and animals and grouping similar organisms together. Biologists still use a modification of Linnaeus's taxonomy today.

All living organisms can be classified as either eukaryotic or prokaryotic. **Eukaryotes** are organisms whose cells contain a nucleus composed of genetic material surrounded by a distinct membrane. **Prokaryotes** are unicellular microbes that lack a true nucleus. Within these categories, microorganisms are further classified as follows:

Bacteria are prokaryotes whose cell walls are composed of peptidoglycan (however, some bacteria lack cell walls). Most are beneficial, but some cause disease.

Archaea are prokaryotes whose cell walls lack peptidoglycan and instead are composed of other chemicals.

Fungi are relatively large microscopic eukaryotes and include **molds** and **yeasts**. These organisms obtain their food from other organisms and have cell walls.

Protozoa are single-celled eukaryotes that are similar to animals in their nutritional needs and cellular structure. Most are capable of locomotion, and some cause disease.

Algae are plantlike eukaryotes that are photosynthetic; that is, they make their own food from carbon dioxide and water using energy from sunlight.

Viruses are microorganisms so small that they were hidden from microbiologists until the invention of the electron microscope in 1932. All are acellular obligate parasites.

Microbiologists also study parasitic worms, which range in size from microscopic forms to adult tapeworms several meters in length.

The Golden Age of Microbiology (pp. 7–18)

During what is sometimes called the "Golden Age" of microbiology, from the late 19th to the early 20th century, microbiologists competed to be the first to answer several questions about the nature of microbial life.

Does Microbial Life Spontaneously Generate?

The theory of **spontaneous generation** (or *abiogenesis*) proposes that living organisms can arise from nonliving matter. It was proposed by Aristotle (384–322 b.c.) and was widely accepted for over 2000 years, until experiments by Francesco Redi (1626–1697) challenged it. In the 18th century, British scientist John T. Needham (1713–1781) conducted experiments suggesting that perhaps spontaneous generation of microscopic life was indeed possible, but in 1799, experiments by Italian scientist Lazzaro Spallanzani (1729–1799) reported results that contradicted Needham's findings. The debate continued until experiments by French scientist Louis Pasteur (1822–1895), using swan-necked flasks that remained free of microbes, disproved the theory definitively.

The debate over spontaneous generation led in part to the development of a generalized **scientific method** by which questions are answered through observations of the outcomes of carefully controlled experiments. It consists of four steps:

1. A group of observations leads a scientist to ask a question about some phenomenon.
2. The scientist generates a hypothesis—a potential answer to the question.
3. The scientist designs and conducts an experiment to test the hypothesis.
4. Based on the observed results of the experiment, the scientist either accepts, rejects, or modifies the hypothesis.

What Causes Fermentation?

The mid-19th century also saw the birth of the field of **industrial microbiology** (or **biotechnology**), in which microbes are intentionally manipulated to manufacture products. Pasteur's investigations into the cause of fermentation led to the discovery that yeast can grow with or without oxygen, and that bacteria ferment grape juice to produce acids, whereas yeast cells ferment grape juice to produce alcohol. These discoveries suggested a method to prevent the spoilage of wine by heating the grape juice just enough to kill contaminating bacteria, so that it could then be inoculated with yeast. *Pasteurization*, the use of heat to kill pathogens and reduce the number of spoilage microorganisms in food and beverages, is an industrial application widely used today.

In 1897, experiments by the German scientist Eduard Buchner (1860–1917) demonstrated the presence of enzymes, cell-produced proteins that promote chemical reactions such as fermentation. His work began the field of **biochemistry** and the study of **metabolism**, a term that refers to the sum of all chemical reactions in an organism.

What Causes Disease?

Prior to the 1800s, disease was attributed to various factors such as evil spirits, astrological signs, imbalances in body fluids, and foul vapors. Pasteur's discovery that bacteria are responsible for spoiling wine led to his hypothesis in 1857 that microorganisms are also responsible for diseases, an idea that came to be known as the **germ theory of disease**. Microorganisms that cause specific diseases are called **pathogens**. Today we know that diseases are also caused by genetics, environmental toxins, and allergic reactions; thus, the germ theory applies only to *infectious* disease.

Investigations into **etiology**, the study of the causation of disease, were dominated by German physician Robert Koch (1843–1910). Koch initiated careful microbiological laboratory techniques in his search for disease agents, such as the bacterium responsible for anthrax. He and his colleagues were responsible for developing techniques to isolate bacteria, stain cells, estimate population size, sterilize growth media, and transfer bacteria between media. They also achieved the first photomicrograph of bacteria. But one of Koch's greatest achievements was the elaboration, in his publications on tuberculosis, of a set of steps that must be taken to prove the cause of any infectious disease. These four steps are now known as **Koch's postulates:**

1. The suspected causative agent must be found in every case of the disease and be absent from healthy hosts.
2. The agent must be isolated and grown outside the host.
3. When the agent is introduced to a healthy, susceptible host, the host must get the disease.
4. The same agent must be found in the diseased experimental host.

In 1884, Danish scientist Christian Gram (1853–1938) developed a staining technique involving application of a series of dyes that leave some microbes purple and others pink. The **Gram stain** is still the most widely used staining technique; it distinguishes Gram-positive from Gram-negative bacteria and reflects differences in composition of the bacterial cell wall.

How Can We Prevent Infection and Disease?

In the mid-19th century, modern principles of hygiene, such as those involving sewage and water treatment, personal cleanliness, and pest control, were not widely practiced. Medical facilities and personnel lacked adequate cleanliness, and *nosocomial infections*, those acquired in a health care facility, were rampant. In approximately 1848, Viennese physician Ignaz Semmelweis (1818–1865) noticed that women whose births were attended by medical students died at a rate 20 times higher than those whose births were attended by midwives in an adjoining wing of the same hospital. Semmelweis hypothesized that "cadaver particles" from the hands of the medical students caused puerperal fever, and he required medical students to wash their hands in chlorinated lime water before attending births. Mortality from puerperal fever in the subsequent year dropped precipitously.

A few years later, English physician Joseph Lister (1827–1912) advanced the idea of antisepsis in health care settings, reducing deaths among his patients by two-thirds with the use of phenol. Florence Nightingale (1820–1910), the founder of modern nursing, introduced antiseptic techniques that saved the lives of innumerable soldiers during the Crimean War of 1854–1856. In 1854, observations by the English physician John Snow (1813–1858) mapping the occurrence of cholera cases in London led to the foundation of two branches of microbiology: **infection control** and **epidemiology**, the study of the occurrence, distribution, and spread of disease in humans.

The field of **immunology**, the study of the body's specific defenses against pathogens, began with the experiments of English physician Edward Jenner (1749–1823), who showed that vaccination with pus collected from cowpox lesions prevented smallpox. Pasteur later capitalized on Jenner's work to develop successful vaccines against fowl cholera, anthrax, and rabies.

The field of **chemotherapy**, a branch of medical microbiology in which chemicals are studied for their potential to destroy pathogenic microorganisms, began when German microbiologist Paul Ehrlich (1854–1915) began to search for a "magic bullet" that could kill microorganisms but remain nontoxic to humans. By 1908, he had discovered chemicals effective against the agents that cause sleeping sickness and syphilis.

The Modern Age of Microbiology (pp. 18–21)

Since the early 20th century, microbiologists have worked to answer new questions in new fields of science.

What Are the Basic Chemical Reactions of Life?

Biochemistry is the study of metabolism. It began with Pasteur's work on fermentation and Buchner's discovery of enzymes, but was greatly advanced by the proposition of microbiologists Albert Kluyver (1888–1956) and C. B. van Niel (1897–1985) that biochemical reactions are shared by all living things, are few in number, and involve the transfer of electrons and hydrogen ions. In adopting this view, scientists could begin to use microbes as model systems to answer questions about metabolism in other organisms. Today, biochemical research has many practical applications, including the design of herbicides and pesticides, diagnosis of illness, treatment of metabolic diseases, and the design of drugs to treat various disorders.

How Do Genes Work?

Microbial genetics is the study of inheritance in microorganisms. Throughout the 20th century, researchers working with microbes made significant advances in our understanding of how genes work. For example, they established that a gene's activity is related to the function of the specific protein coded by that gene, and they determined the exact way in which genetic information is translated into a protein.

Molecular biology combines aspects of biochemistry, cell biology, and genetics to explain cell function at the molecular level. It is particularly concerned with genome sequencing. **Genetic engineering** involves the manipulation of genes in microbes, plants, and animals for practical applications, such as the development of pest-resistant crops and the treatment of disease. **Gene therapy** is the use of recombinant DNA (DNA composed of genes from more than one organism) to insert a missing gene or repair a defective gene in human cells.

What Roles Do Microorganisms Play in the Environment?

Environmental microbiology studies the role microorganisms play in their natural environment. Microbial communities play an essential role, for example, in the decay of dead organisms and the recycling of chemicals such as carbon, nitrogen, and sulfur. Environmental microbiologists study the microbes and chemical reactions involved in such biodegradation, as well as the effects of community-based measures to limit the abundance of pathogenic microbes in the environment, such as sewage treatment, water purification, and sanitation measures.

How Do We Defend Against Disease?

Although the work of Jenner and Pasteur marked the birth of the field of immunology, the discovery of chemicals in the blood that are active against specific pathogens advanced the field considerably. *Serology* is the study of blood serum, the liquid that remains after blood coagulates and carries disease-fighting chemicals. Serologic studies showed that the body can defend itself against a remarkable range of diseases. Nevertheless, medical intervention is often necessary, and the 20th century saw tremendous advances in chemotherapy, including the discovery of penicillin in 1929 and sulfa drugs in 1935, both of which are still first-line antimicrobial drugs today.

What Will the Future Hold?

Among the questions microbiologists are working to answer today are the following:

What prevents certain life forms from being grown in the laboratory?

Can microorganisms be used in ultraminiature technologies, such as computer circuit boards?

How can an understanding of microbial communities help us understand the positive aspects of microbial action?

How can we reduce the threat from microbes resistant to antimicrobial drugs and conquer reemerging infectious diseases?

KEY THEMES

Though it might not seem as important as microbial structure or metabolism, history is fundamental to the story of microbiology. Some of the most important techniques used today to analyze and identify microbes have very old historical roots. The questions of yesterday in many cases remain unanswered, awaiting future resolution. As you study the material in Chapter 1, focus on the following key ideas:

The central tenets of the scientific method: Early experiments by Pasteur, Koch, and others helped to lay the foundations of the scientific method. By learning and understanding the scientific method, you will be better prepared to analyze the complex problems presented later in the text and solve them.

Themes of microbiological research: The Modern Age of microbiology is deeply rooted in old questions about the relationships between microbes and between microbes and humans. By looking at the questions posed during the Golden Age of microbiology, we can discern the patterns of microbial research that allow us to decipher these relationships.

QUESTIONS FOR FURTHER REVIEW

Answers to these questions can be found in the answer section at the back of this study guide. Refer to the answers only after you have attempted to solve the questions on your own.

Multiple Choice

1. Leeuwenhoek's most important contribution to the field of microbiology was
   a. the first microscope.
   b. the first description of viruses.
   c. the first description of diseases caused by microbes.
   d. the first technique for growing microbes in the laboratory.

2. Which microbial group was NOT discovered until the introduction of modern research tools?
   a. archaea
   b. bacteria
   c. fungi
   d. viruses

3. Who was the first scientist to challenge the theory of spontaneous generation?
   a. John T. Needham
   b. Louis Pasteur
   c. Francesco Redi
   d. Lazzaro Spallanzani

4. Early scientists used the term *fermentation* to mean
   a. the formation of alcohol.
   b. the formation of acids.
   c. the putrefaction of meat.
   d. all of the above.

5. The germ theory of disease specifically applies to
   a. all diseases.
   b. genetic diseases.
   c. infectious diseases.
   d. diseases caused by toxins or allergies.

6. What laboratory technique became important as a means to satisfy Koch's first postulate?
   a. isolation of organisms in pure culture
   b. observation of organisms using microscopy
   c. staining techniques, such as the Gram stain
   d. methods for counting the numbers of organisms present in a sample.

7. The Gram stain is useful for distinguishing among which types of microbes?
   a. algae
   b. bacteria
   c. fungi
   d. protozoa

8. Who was the first physician to truly gain acceptance for the concept of antisepsis?
   a. Edward Jenner
   b. Joseph Lister
   c. Ignaz Semmelweis
   d. John Snow

9. Which of the following questions was NOT answered during the Golden Age of microbiology?
   a. Does life spontaneously generate?
   b. Do microbes cause fermentation?
   c. Does microbial life exist on other planets?
   d. Do microbes cause disease?

10. Which of the following is NOT a role that microbes play in the environment?
   a. causing disease
   b. decaying dead organisms
   c. recycling chemicals
   d. All are roles played by microbes.

Fill in the Blanks

1. Leeuwenhoek's "beasties" are now often called ____________________.

2. Another term for spontaneous generation is ____________________.

3. A specific microbe that is responsible for producing a specific disease is called a ____________________.

4. Infections acquired in health care settings as a result of inadequate cleanliness are called nosocomial infections.

5. The first physician to use a vaccine on another individual was Ed. Jenner.

Short-Answer Questions for Thought and Review

1. Describe the fundamental differences between viruses and other prokaryotic and eukaryotic microbes.

   Dependant on host.

2. Compare the leading questions posed during the Golden Age of microbiology with the leading questions of today. What has changed with respect to what we want to know? What questions remain from the Golden Age for scientists to answer?

   Spont. Gen? No!
   Microbes function? ✓
   Planets

3. Explain the concept of control groups and their significance to the scientific method.

4. Florence Nightingale documented that poor food and unsanitary conditions contributed to the deaths of soldiers in field hospitals. How might such conditions influence death?

Critical Thinking

1. Carolus Linnaeus developed the first system for grouping similar organisms together. This system has since been greatly modified, but the goal of categorizing organisms based on genetic and evolutionary relationships remains. How might these deduced relationships between microbes better help microbiologists understand the relationship between microbes and the diseases they cause?
2. Ignaz Semmelweis was one of the first physicians to propose handwashing as a means of preventing the spread of disease. His observations are described in Chapter 1. The scientific method, also described in Chapter 1, came into being at about the same time. Ethically, Semmelweis could not have followed the scientific method to investigate his hypotheses. Why not?
3. Paul Ehrlich dedicated his scientific career to searching for "magic bullets" to cure infectious diseases. Why is it unlikely that a single "magic bullet" will ever be found to cure all infectious diseases?

2 The Chemistry of Microbiology

CHAPTER SUMMARY

Atoms (pp. 27–29)

Matter is defined as anything that takes up space and has mass. The smallest chemical units of matter are **atoms**.

Atomic Structure

Atoms contain negatively charged particles called **electrons** spinning around a nucleus composed of uncharged particles called **neutrons** and positively charged particles called **protons**. (A hydrogen atom contains only one proton and no neutrons.) The number of electrons in an atom typically equals the number of protons, so atoms are electrically neutral overall.

An **element** is matter that is composed of a single type of atom. Of the 93 naturally occurring elements, organisms utilize only about 20, including, for example, carbon, oxygen, and nitrogen. Elements differ from one another in their **atomic number**, which is the number of protons in their nuclei. The **atomic mass** (or atomic weight) of an atom is the sum of the masses of its protons, neutrons, and electrons. Because electrons have little mass, the atomic mass is estimated by adding the number of protons and neutrons. Thus, hydrogen has an atomic mass of 1.

Isotopes

Isotopes are atoms of an element that differ only in the numbers of neutrons they contain. For example, there are three naturally occurring isotopes of carbon, all of which have six protons: carbon-12 has six neutrons, carbon-13 has seven neutrons, and carbon-14 has eight neutrons.

Electron Configurations

Because only the electrons of atoms come close enough to interact, they determine an atom's chemical behavior. Electrons orbit their nucleus in three-dimensional electron shells, each of which can hold only a certain maximum number of electrons. For example, the first shell of any atom has a capacity of just two electrons, whereas the second shell has a capacity of eight. The number of electrons in the **valence shell**—the outermost shell—determines the atom's reactivity: atoms with valence shells not containing the maximum number of electrons are more likely to give up or accept electrons from another atom until their outermost shell is full.

Chemical Bonds (pp. 29–34)

The sharing or transferring of electrons to fill a valence shell results in the formation of **chemical bonds**. Two or more atoms held together by chemical bonds form a **molecule**. Any molecule containing atoms of more than one element is called a compound. For example, two hydrogen atoms bonded to an oxygen atom form a molecule of water (H_2O), which is a compound.

Nonpolar Covalent Bonds

A **covalent bond** is the sharing of a pair of electrons by two atoms. Two hydrogen atoms bind covalently to form a stable molecule of hydrogen in which both atoms have full valence shells. Atoms such as oxygen that share two pairs of electrons have a double covalent bond with each other. The attraction of an atom for electrons is called its **electronegativity**. When atoms with similar electronegativities bind, the shared electrons tend to spend an equal amount of time around each nucleus of the pair. Because neither nucleus acts as a "pole" to exert an unequal pull, these are called **nonpolar covalent bonds**.

Carbon atoms have four electrons in their valence shells, so they have an equal tendency to lose or gain four electrons, and form nonpolar covalent bonds with one another and with many other atoms. One result of this feature is that carbon atoms easily form very long chains that constitute the "backbone" of many biologically important molecules. Compounds that contain carbon and hydrogen atoms are called **organic compounds**.

Polar Covalent Bonds

When atoms with significantly different electronegativities combine, the electron pair will spend more time orbiting the "pole"—that is, the nucleus of the atom with greater electronegativity. Bonds with an unequal sharing of electrons are therefore called **polar covalent bonds**. A water molecule, for example, has two polar covalent bonds. Although they can form between many different elements, the most biologically important polar covalent bonds are those involving hydrogen because they allow hydrogen bonding.

Ionic Bonds

When two atoms with vastly different electronegativities approach each other, the atom with the higher electronegativity will strip one or more electrons from the valence shell of the other. This happens, for example, when chlorine, with seven electrons in its valence shell, encounters sodium, which has just one valence electron. When sodium loses an electron, it becomes positively charged; when chlorine gains an electron, it becomes negatively charged. Charged atoms are called **ions**; specifically, an ion like sodium with a positive charge is called a **cation**, whereas an atom like chlorine with a negative charge is called an **anion**. The opposite charges of cations and anions attract each other strongly to form an **ionic bond**. Molecules with ionic bonds form crystalline compounds known as **salts**, such as sodium chloride (NaCl). Notice that in ionic bonds, electrons are transferred from one molecule to another; in contrast to covalent bonds, there is no sharing of electrons.

The polar bonds of water molecules interfere with the ionic bonds of salts, causing dissociation (also called ionization). When cations and anions dissociate in water, they are called **electrolytes** because they can conduct electricity through the solution.

Hydrogen Bonds

Like ionic bonds, **hydrogen bonds** do not involve the sharing of electrons. Instead, a partially charged hydrogen atom is attracted to a full or partial negative charge on either a different region of the same molecule or another molecule. The cumulative effect of numerous hydrogen bonds is to stabilize the three-dimensional shapes of large molecules, such as DNA. Thus, although weak, they are essential to life.

Chemical Reactions (pp. 34–36)

Chemical reactions result from the making or breaking of chemical bonds in a process in which **reactants**—the atoms, ions, or molecules that exist at the beginning of a reaction—are changed into **products**—the atoms, ions, or molecules that remain after the reaction is complete.

Synthesis Reactions

Synthesis reactions involve the formation of larger, more complex molecules. An important type is **dehydration synthesis**, in which two smaller molecules are joined together by a covalent bond, and a water molecule is removed from the reactants. Synthesis reactions require energy to break bonds in the reactants and to form new bonds to make products. They are said to be **endothermic reactions** because they trap energy within new molecular bonds. **Anabolism** is the sum of all synthesis reactions in an organism.

Decomposition Reactions

Decomposition reactions are the opposite of synthesis reactions in that they break bonds within larger reactants to form smaller atoms, ions, and molecules. Because these reactions release energy, they are called **exothermic.** A common type of decomposition reaction is **hydrolysis**, the reverse of dehydration synthesis, in which a covalent bond in a large molecule is broken, and the ionic components of water (H^+ and OH^-) are added to the products. Collectively, all the decomposition reactions in an organism are called **catabolism.**

Exchange Reactions

Exchange reactions involve exchanging atoms between reactants. An important example is the phosphorylation of glucose. The sum of all chemical reactions in an organism is called **metabolism.**

Water, Acids, Bases, and Salts (pp. 36–39)

Inorganic chemicals lack carbon. Many, including water, acids, bases, and salts, are essential to life.

Water

Water constitutes 50–99% of the mass of living organisms. It is vital to life because of its solvent properties, its liquidity, its capacity to absorb heat, and its participation in chemical reactions. In addition, water molecules are cohesive; they stick to one another via hydrogen bonding. This property generates surface tension, which enables water to form a thin layer on the surface of cells through which dissolved molecules can be transported into and out of the cell.

Acids and Bases

An **acid** is a substance that dissociates into one or more hydrogen ions (H^+) and one or more anions. A **base** is a molecule that binds with H^+ when dissolved in water. Many bases dissociate into hydroxyl ions and cations. The concentration of hydrogen ions in a solution is expressed using a logarithmic **pH** scale in which acidity increases as pH values decrease. Organisms can tolerate only a narrow pH range. Thus, most organisms contain natural **buffers**, substances that prevent drastic changes in internal pH.

Salts

A **salt** is a compound that dissociates in water into cations and anions other than H^+ and OH^-. A cell uses the cations and anions of salts—electrolytes—to create electrical differences between its internal and external environments, to transfer electrons from one location to another, and as important components of many enzymes. Some organisms use salts to provide structural support for their cells.

Organic Macromolecules (pp. 39–51)

Organic macromolecules are generally large, complex molecules containing carbon and hydrogen atoms linked together in branched and unbranched chains, and rings bound to one or more other elements, such as oxygen, nitrogen, phosphorus, or sulfur.

Functional Groups

In organic macromolecules, atoms often appear in certain common arrangements called **functional groups**. For example, the hydroxyl functional group is common to all alcohols. In addition, the organic macromolecules of proteins, carbohydrates, and nucleic acids are composed of simple subunits called **monomers** that can be covalently linked to form chainlike **polymers**, which may be hundreds of thousands of monomers long.

Lipids

Lipids are organic macromolecules composed almost entirely of carbon and hydrogen atoms linked by nonpolar covalent bonds. Because they are nonpolar, they are **hydrophobic**—that is, they are insoluble in water. There are four major groups of lipids:

> **Fats** are composed of glycerol and three chainlike fatty acids. **Saturated fatty acids** contain more hydrogen in their structural formulas than do **unsaturated fatty acids**, which contain double bonds between some of their carbon atoms. If several double bonds exist, the fatty acid is called a **polyunsaturated fat.**
>
> **Phospholipids** contain only two fatty acid chains and a phosphate functional group. The fatty acid "tail" of a phospholipid molecule is nonpolar and thus hydrophobic; the phospholipid "head" is polar and thus hydrophilic. This means that phospholipids placed in a watery environment will always self-assemble into forms that keep the fatty acid tails away from water, such as is found in the outer membranes of all cells.
>
> **Waxes** contain one long-chain fatty acid linked covalently to a long-chain alcohol by an ester bond. They are completely water insoluble and are sometimes used as energy storage molecules.
>
> **Steroids** consist of four carbon rings that are fused to one another and attached to various side chains and functional groups. Many organisms have sterol molecules in their cell membranes that keep them fluid at low temperatures.

Carbohydrates

Carbohydrates are organic molecules composed solely of atoms of carbon, hydrogen, and oxygen. They are used for immediate and long-term storage of energy, as structural components of DNA and RNA and some cell walls, and for conversion into amino acids. They also serve as recognition sites during intercellular interactions. There are three basic groups:

Monosaccharides are simple sugars, such as glucose and fructose. They usually take cyclic forms.

Disaccharides are formed when two monosaccharides are linked together via dehydration synthesis. Sucrose, lactose, and maltose are examples.

Polysaccharides are polymers composed of tens, hundreds, or thousands of monosaccharides that have been covalently linked in dehydration synthesis reactions.

Cellulose and glycogen are examples.

Proteins

The most complex organic macromolecules are **proteins**, which are composed mostly of carbon, hydrogen, oxygen, nitrogen, and sulfur. They function as structural components of cells, enzymatic catalysts, regulators of various activities, transporters of substances, and defense molecules. The monomers of proteins are **amino acids**, in which a central carbon is attached to an amino group, a hydrogen atom, a carboxyl group, and a side group that varies according to the amino acid. These are linked by **peptide bonds** into specific structural patterns determined genetically. Every protein has at least three levels of structure, and some have four. **Denaturation** of a protein disrupts its structure and consequently its function.

Nucleic Acids

The two nucleic acids **deoxyribonucleic acid (DNA)** and **ribonucleic acid (RNA)** comprise the genetic material of cells and viruses. These differ primarily in the structure of their monomers, which are called nucleotides. Each nucleotide consists of phosphate, a pentose sugar (deoxyribose or ribose), and one of five cyclic nitrogenous bases: **adenine (A), guanine (G), cytosine (C), thymine (T), or uracil (U).** DNA contains A, G, C, and T nucleotides, whereas RNA contains A, G, C, and U nucleotides.

The structure of nucleic acids allows for genetic diversity, correct copying of genes for their passage on to the next generation, and accurate synthesis of proteins. **Adenosine triphosphate** (ATP), which is made up of the nitrogenous base adenine, ribose sugar, and three phosphate groups, is the most important short-term energy storage molecule in cells. It is also incorporated into the structure of many coenzymes.

KEY THEMES

Many biology (and microbiology) students want to know why they have to know so much chemistry. The answer is relatively simple: chemistry is the basis for all life on Earth. As you study this chapter, keep in mind the following:

Chemical bonds are the foundation of all relationships: Chemical bonds determine how atoms, molecules, compounds, and cells interact with each other physically. It is important to know the different types of bonds possible in nature to firmly understand these relationships engendered by these physical associations.

Chemical reactions are central to life: Chemical reactions occur because of the existence of chemical bonds. Nature's ability to form and reform these bonds sustains life in a process called metabolism. To understand the very basis of our survival, and the survival of all other living creatures on Earth, we need to understand the essential elements of chemistry.

QUESTIONS FOR FURTHER REVIEW

Answers to these questions can be found in the answer section at the back of this study guide. Refer to the answers only after you have attempted to solve the questions on your own.

Multiple Choice

1. Which element has the greatest number of protons in its nucleus?
   a. carbon-12
   b. carbon-14
   c. carbon-13
   d. All have the same number of protons.
2. A nonpolar covalent bond
   a. shares electrons equally between atoms.
   b. shares electrons unequally between atoms.
   c. forms when one atom strips the valence electrons from another atom.
   d. does not form between the electrons of different atoms.
3. Which of the following acids is an organic acid?
   a. HNO_3 (nitric acid)
   b. HCl (hydrochloric acid)
   c. CH_3COOH (acetic acid)
   d. H_3PO_4 (phosphoric acid)
4. Salts are held together by what types of bonds?
   a. nonpolar covalent bonds
   b. polar covalent bonds
   c. ionic bonds
   d. hydrogen bonds
5. Exchange, or transfer, reactions are
   a. endothermic.
   b. exothermic.
   c. both endothermic and exothermic.
   d. neither endothermic nor exothermic.
6. Lemon juice, with a pH of 2, contains how many more hydrogen atoms than milk, with a pH of 6?
   a. 10 times
   b. 100 times
   c. 1,000 times
   d. 10,000 times
7. Which of the following is NOT considered a macromolecule?
   a. DNA
   b. amino acid
   c. phospholipids
   d. antibodies
8. Lipids are composed almost entirely of carbon and hydrogen atoms linked together by
   a. nonpolar covalent bonds.
   b. polar covalent bonds.
   c. ionic bonds.
   d. hydrogen bonds.

9. What type of fatty acids would you expect to see predominating in the membranes of microbes living in extremely hot environments?
   a. saturated fatty acids
   b. unsaturated fatty acids
   c. polyunsaturated fatty acids
   d. All types of fatty acids would be present in equal amounts.
10. The cell walls of bacteria are made out of peptidoglycan composed of what materials?
    a. monosaccharides and amino acids
    b. polysaccharides and amino acids
    c. monosaccharides and lipids
    d. polysaccharides and lipids
11. What is the primary energy molecule of cells?
    a. amylose
    b. cellulose
    c. glucose
    d. glycogen
12. Which of the following is NOT a function of nucleic acids?
    a. carrying the genetic instructions of the cell
    b. involvement in protein synthesis
    c. providing energy for cellular functions
    d. None of these are functions of nucleic acids.

Fill in the Blanks

1. The atomic number of an element corresponds to the number of ____________________ in its nucleus.
2. Chemical bonds form between two or more ____________________ to create a ____________________.
3. Two ways to write the structural formula for O_2 are ____________________ and ____________________.
4. For each of the following reactions, indicate the type of reaction that is occurring:
   a. $C_6H_{12}O_6 + 6O_2 \rightarrow 6H_2O + 6CO_2$ ____________________
   b. $C_6H_{12}O_6 + ATP \rightarrow C_6H_{12}O_6{-}P + ADP$ ____________________
   c. Glucose + Fructose → sucrose + H_2O ____________________
5. Acids characteristically dissociate to release ____________________ ions. Bases dissociate to release ____________________ ions. (Give the name and the molecular formula for each.)

6. Organisms cannot tolerate large changes in internal pH and so they contain natural ____________________ to protect themselves. An example is ____________________.

7. Phospholipids form membranes by orienting their ____________________ head groups toward the water environment and their ____________________ tails away from the water.

8. Covalent bonds formed between amino acids in a protein are called ____________________ bonds. Other bonds, mostly ____________________ bonds, help to hold the protein in its tertiary form.

9. In DNA, the pentose sugar that forms each nucleotide is ____________________, whereas in RNA the sugar is ____________________.

10. Write the complementary DNA strand sequence and the complementary RNA strand sequence for the following sequence: 5' - ATTGCTACCGAT - 3'.

    a. DNA sequence: ____________________.

    b. RNA sequence: ____________________.

Short-Answer Questions for Thought and Review

1. Describe the difference between an element, a molecule, and a compound. Give one example of each.

2. Using a periodic table, as shown in Figure 2.4 of the textbook, draw the electron shells of calcium and chlorine. Indicate the valence for each element. Draw the molecule $CaCl_2$, indicating the type of bonds formed.

3. Fats are believed to contribute to certain types of heart disease, such as atherosclerosis, by building up in the arteries of the heart and blocking them. Which types of fatty acids—saturated, unsaturated, or polyunsaturated—would you expect to contribute least to heart disease, and why?

4. List the five primary functions of proteins. For each function, explain in one sentence what would happen to an organism if proteins could no longer perform that function.

Critical Thinking

1. In metabolism, catabolism provides the energy necessary for anabolism to occur. Consider a bacterium living in the human intestinal tract versus one living in the water of an abandoned flowerpot. Which one would you predict to have the higher metabolic activity, and why? Define the terms metabolism, catabolism, and anabolism in your answer.
2. Whereas microbes require a relatively neutral internal pH, many species can survive and even thrive in highly acidic or basic external environments. Some microbes even physically alter the pH of their environment to better suit themselves. How does this ability to change the acidic or basic nature of the environment confer a growth advantage on the microbes that do it?
3. Mutations, in the form of amino acid substitutions, happen at a relatively high frequency in microbes. Some are lethal and destroy protein function, whereas others are not. These neutral, nonharmful substitutions are generally believed to predominate in nature. Explain why this might be so.

Concept Building Questions

1. One of the key questions posed during the Golden Age of microbiology was what caused fermentation to occur (see Chapter 1). What type(s) of chemical reactions is(are) occurring in the microbial production of alcohol from sugar? In general, why are these types of reactions useful for the cell?
2. The role microbes play in the environment is an important question to microbiologists. Reread the section in Chapter 1 that describes this question, and then explain how a firm understanding of chemical bonds and chemical reactions is crucial to answering this question.

3 Cell Structure and Function

CHAPTER SUMMARY

Processes of Life (pp. 57–58)

All living things share four processes:

Growth: an increase in size
Reproduction: an increase in number
Responsiveness: an ability to react to environmental stimuli
Metabolism: controlled chemical reactions

In addition, all living organisms share a cellular structure. Although viruses have some characteristics of living cells, they cannot grow, and they reproduce only when inside a host cell. They also depend on a host cell's metabolism, and they have no cellular structure. For these reasons, microbiologists debate the question of whether viruses are truly alive.

Prokaryotic and Eukaryotic Cells: An Overview (pp. 58–59)

Cells can be classified as prokaryotic or eukaryotic. **Prokaryotic** cells, such as bacteria and archaea, lack a nucleus and membrane-bound organelles. **Eukaryotic** cells, such as the cells of animals, plants, algae, fungi, and protozoa, have internal, membrane-bound organelles, including true nuclei. Prokaryotic and eukaryotic cells have some common structural features such as external structures, cell walls, cytoplasmic membranes, and cytoplasm.

External Structures of Bacterial Cells (pp. 59–65)

The external structures of bacterial cells include glycocalyces, flagella, fimbriae, and pili.

Glycocalyces

A **glycocalyx** is a gelatinous, sticky substance that surrounds the outside of the cell. When the glycocalyx of a bacterium is firmly attached to the cell surface, it is called a **capsule**. When loose and water soluble, it is called a **slime layer**. Both types protect the cell from desiccation, and both increase the cell's ability to cause disease. Capsules protect cells from phagocytosis, and slime layers enable cells to adhere to each other and to environmental surfaces.

Flagella

The structures responsible for cell motility include **flagella**, long extensions from the cell surface and glycocalyx that propel a cell through its environment. Bacterial flagella are composed of a filament, a hook, and a basal body. Flagella covering the cell are termed peritrichous flagella, and those found at the ends of a cell are called polar flagella.

Endoflagella are the special flagella of spirochetes that spiral tightly around the cell instead of protruding into the environment. Together, these endoflagella form an **axial filament** that wraps around the cell and rotates, enabling it to "corkscrew" through its medium.

Flagella enable bacterial cells to move clockwise or counterclockwise, in a series of runs and tumbles. Via **taxis**, flagella move the cell toward or away from stimuli such as chemicals (**chemotaxis**) or light (**phototaxis**).

Fimbriae and Pili

Fimbriae are short, sticky, proteinaceous extensions of some bacteria that help cells adhere to one another and to substances in the environment. They serve an important function in **biofilms**, slimy masses of microbes adhering to a surface.

Pili (also called **conjugation pili**) are tubules of a protein called pilin that connect some prokaryotic cells. Typically, only one or a few are present per cell. They join two bacterial cells and mediate the movement of DNA from one cell to another, a process called conjugation.

Bacterial Cell Walls (pp. 65–68)

Most prokaryotic cells are surrounded by a **cell wall** (not found in eukaryotes) that provides structure and protection from osmotic forces. A few bacteria lack cell walls entirely, but most have walls composed of **peptidoglycan,** a complex polysaccharide composed of two alternating sugars called *N-acetylglucosamine (NAG)* and *N-acetylmuramic acid (NAM)*. Chains of NAG and NAM are attached to other chains by crossbridges of four amino acids (tetrapeptides).

Gram-Positive Bacterial Cell Walls

Gram-positive bacterial cells have thick layers of peptidoglycan that also contain teichoic acids. Their thick wall retains the crystal violet dye used in the Gram staining procedure, so the stained cells appear purple under magnification.

Gram-Negative Bacterial Cell Walls

Gram-negative bacterial cells have only a thin layer of peptidoglycan, outside which is a membrane containing **lipopolysaccharide (LPS)**. LPS is composed of sugars and a lipid known as lipid A. During an infection with Gram-negative bacteria, as the outer membrane of dead cells disintegrate, lipid A accumulates in the blood and may cause shock, fever, and blood clotting. Between the cytoplasmic membrane and the outer membrane is a **periplasmic space** containing peptidoglycan. Because the cell walls of Gram-negative bacteria differ from those of Gram-positive bacteria, Gram-negative cells appear pink.

Bacteria without Cell Walls

A few bacteria, such as *Mycoplasma pneumoniae*, lack cell walls entriely. However, they still possess the other features of prokaryotic cells, such as prokaryotic ribosomes.

Bacterial Cytoplasmic Membranes (pp. 68–73)

Beneath the glycocalyx and cell wall is a **cytoplasmic membrane** (or cell membrane).

Structure

The cytoplasmic membrane is a double-layered structure, called a **phospholipid bilayer**, composed of molecules with hydrophobic lipid tails and hydrophilic phosphate heads. Proteins associated with the membrane vary in location and function and are able to flow laterally within it. The **fluid mosaic model** is descriptive of the current understanding of the membrane. Archaea do not have phospholipid membranes, and some have a single layer of lipid instead of a bilayer.

Function

The **selectively permeable** cytoplasmic membrane not only separates the contents of the cell from the outside environment but also controls the contents of the cell, allowing some substances in and out while preventing the movement of others. Although impermeable to most substances, its proteins act as pores, channels, or carriers to allow or facilitate the transport of substances the cell needs. The relative concentration of chemicals (**concentration gradients**) inside and outside the cell and of the corresponding electrical charges, or voltage (**electrical gradients**) create an overall electrochemical gradient across the membrane. A cytoplasmic membrane uses the energy inherent in its electrochemical gradient to transport substances into or out of the cell.

Passive Processes

Passive processes require no energy expenditure to move chemicals across the cytoplasmic membrane. **Diffusion** is the movement of chemicals down their concentration gradient, from an area of higher concentration to an area of lower concentration. In **facilitated diffusion**, proteins act as channels or carriers to allow certain molecules to diffuse into or out of the cell along their electrochemical gradient. **Osmosis** is the diffusion of water molecules across a selectively permeable membrane in response to differing concentrations of solutes. Concentrations of solutes can be compared as follows: **hypertonic** solutions have a higher concentration of solutes than **hypotonic** solutions. For example, seawater is hypertonic to distilled water. Two **isotonic** solutions have the same concentration of solutes.

Active Processes

Active processes require cells to expend energy in the form of ATP to move chemicals across the cytoplasmic membrane against their concentration gradient. **Active transport** moves substances via transmembrane permease proteins, which may transport two substances in the same direction at once (symports) or move substances in opposite directions (antiports). **Group translocation**, which occurs only in some bacteria, causes chemical changes to the substance being transported. The membrane is impermeable to the altered substance, which is then trapped inside the cell. One well-studied example is the phosphorylation of glucose.

Cytoplasm of Bacteria (pp. 73–76)

Cytoplasm is the gelatinous, elastic material inside a cell. It is composed of cytosol, inclusions, ribosomes, and, in many cells, a cytoskeleton. Some bacterial cells produce internal, resistant, dormant forms called endospores.

Cytosol

The liquid portion of the cytoplasm is called **cytosol**. It is mostly water, plus dissolved and suspended substances such as ions, carbohydrates, proteins, lipids, and wastes. The cytosol of prokaryotes also contains the cell's DNA in a region called the nucleoid.

Inclusions

Deposits called **inclusions** may be found within bacterial cytosol. These may be reserve deposits of lipids, starch, or other chemicals. Inclusions called gas vesicles store gases.

Endospores

Some bacteria produce structures called **endospores** when one or more nutrients are limited. Endospores can survive under harsh conditions, making them a concern to food processors and health care professionals.

Nonmembranous Organelles

Two types of **nonmembranous organelles** are found in direct contact with the cytosol in bacteria: ribosomes and the cytoskeleton. **Ribosomes** are the sites of protein synthesis in cells. They are composed of protein and ribosomal RNA (rRNA). Their size is expressed in Svedbergs (S) and is determined by their sedimentation rate: bacterial ribosomes are 70S and are smaller than 80S eukaryotic ribosomes. The **cytoskeleton** is an internal network of fibers that plays a role in forming a cell's basic shape. Once thought to lack cytoskeletons, bacteria have been shown to have simple ones.

External Structures of Archaea (pp. 76–77)

Glycocalyces

Like the glycocalyces of bacteria, **glycocalyces of archaea** are composed of polysaccharides, polypeptides, or both. Archaeal glycocalyces function in forming biofilms by adhering cells to one another, to other types of cells, and to inanimate surfaces in the environment. Whereas bacterial biofilms are associated with disease, no link has been demonstrated between archaeal glycocalyces or biofilms and disease.

Flagella

Archaea use **flagella** to move through the environment. Archaeal flagella are superficially similar to bacterial flagella, and consist of a basal body, hook, and filament. Archaeal flagella are different from bacterial flagella in that they are thinner, have proteins with distinct amino acid sequences compared to the proteins of bacterial flagella, and are powered by energy stored in ATP.

Fimbriae and Hami

Many archaea have **fimbriae** that anchor the cells to one another and to environmental surfaces. Some archaea have unique fimbriae-like structures called **hami**. Hami also function to attach archaea to biological and inanimate surfaces.

Archaeal Cell Walls and Cytoplasmic Membranes (p. 77–78)

Most archaea have cell walls, although they lack peptidoglycan, a component of bacterial cell walls. Like bacteria, archaea can be grouped as Gram-positive or Gram-negative based on their cell wall components. The cytoplasmic membrane of archaea functions similarly to that of bacteria, maintaining electrical and chemical gradients and controlling the movement of substances into and out of the cell.

Cytoplasm of Archaea (pp. 78)

The **cytoplasm of archaea** has features of the cytoplasm of both bacterial and eukaryotic cells. Archaea, like bacteria, do not have membranous organelles in their cytoplasm. Like bacteria, archaea have 70S ribosomes, but archaeal ribosomes are composed of proteins more similar to those found in eukaryotic ribosomes. The enzymes archaea use to make RNA are different from those used by bacteria. Also, the genetic code of archaea is more like that of eukaryotes.

External Structures of Eukaryotic Cells (pp. 78)

Eukaryotic cells have many external structures similar to those of prokaryotes, as well as some unique features.

Glycocalyces

Glycocalyces are absent in eukaryotes with cell walls, but animal and protozoan cells—which lack cell walls—do have glycocalyces anchored to their cytoplasmic membranes. They strengthen the cell surface, provide protection against dehydration, and function in cell-to-cell recognition and communication.

Eukaryotic Cell Walls and Cytoplasmic Membranes (pp. 79–81)

The eukaryotic cells of fungi, algae, plants, and some protozoa lack glycocalyces; instead, a cell wall composed of polysaccharides provides protection from the environment. It also provides shape and support against osmotic pressure. The cell walls of plants are composed of cellulose, whereas fungal cell walls are composed of chitin or other polysaccharides. Algal cell walls contain agar, carrageenan, algin, or other chemicals.

All eukaryotic cells have **cytoplasmic membranes**. Like bacterial membranes, they are a fluid mosaic of phospholipids and proteins. Unlike bacterial membranes, they contain steroids that strengthen and solidify the membrane when temperatures rise and help maintain fluidity when temperatures fall. Some eukaryotic cells transport substances into the cytoplasm via **endocytosis**, which is an active process requiring the expenditure of energy by the cell. In endocytosis, **pseudopodia**—movable extensions of the cytoplasm and membrane of the cell—surround a substance and move it into the cell. When solids are brought into the cell, endocytosis is called **phagocytosis**. The incorporation of liquids is called pinocytosis.

Cytoplasm of Eukaryotes (pp. 81–89)

The **cytoplasm** of eukaryotes is more complex than that of prokaryotes, containing a few nonmembranous and numerous membranous organelles.

Flagella

Eukaryotic flagella are within the cytoplasmic membrane. The shaft of a eukaryotic flagellum is composed of molecules of a globular protein called **tubulin** arranged in chains to form hollow microtubules arranged in nine pairs around a central two microtubules. The basal body also has microtubules, but in triplets with no central pair. Eukaryotic flagella have no hook and do not extend outside the cell. Rather than rotating, eukaryotic flagella undulate rhythmically to push or pull the cell through the medium. They exhibit taxis.

Cilia

Some eukaryotic cells are covered with **cilia**, which have the same structure as eukaryotic flagella but are much shorter and more numerous. Their rhythmic beating propels single-celled eukaryotes through their environment. Some more complex organisms use cilia to sweep substances in the local environment, such as dust particles, past the surface of the cell.

Other Nonmembranous Organelles

Three nonmembranous organelles are found in eukaryotes: **ribosomes, a cytoskeleton, and centrioles.** Eukaryotic ribosomes are 80S and are found within the cytosol as well as attached to the membranes of the endoplasmic reticulum, discussed below. The cytoskeleton is extensive and is composed of both fibers and tubules. It acts to anchor organelles and functions in cytoplasmic streaming and movement of organelles within the cytosol. Cytoskeletons in some cells enable the cell to contract, move the cytoplasmic membrane during endocytosis and amoeboid action, and produce the basic shapes of many cells. In addition, animal cells and some fungal cells contain two centrioles lying at right angles to each other near the nucleus, in a region of the cytoplasm called the centrosome. Centrioles are composed of nine triplets of tubulin microtubules. Centrosomes play a role in mitosis and cytokinesis and in the formation of flagella and cilia.

Membranous Organelles

Eukaryotic cells contain a variety of organelles that are surrounded by **phospholipid bilayer membranes** similar to the cytoplasmic membrane.

Nucleus

The **nucleus** is spherical to ovoid and is often the largest organelle in a cell. It contains most of the cell's genetic material in the form of DNA. The semiliquid matrix of the nucleus is called the **nucleoplasm.** Within it, one or two specialized regions of RNA synthesis, called **nucleoli,** may be present. The nucleoplasm also contains **chromatin,** a threadlike mass of DNA and associated histone proteins. Chromatin becomes visible as chromosomes during mitosis (Chapter 12). Surrounding the nucleus is a double membrane called the **nuclear envelope,** which contains **nuclear pores** that function to control the import and export of substances through the envelope.

Endoplasmic Reticulum

Continuous with the outer membrane of the nuclear envelope and traversing the cytoplasm is a net of flattened hollow tubules called endoplasmic reticulum (ER). Smooth endoplasmic reticulum (SER) plays a role in lipid synthesis and transport.

Ribosomes adhere to the surface of rough endoplasmic reticulum (RER) and produce proteins that are transported throughout the cell.

Golgi Body

The **Golgi body** is a series of flattened, hollow sac surrounded by phospholipid bilayers. It receives, processes, and packages large molecules in secretory vesicles, which release their contents from the cell via exocytosis.

Lysosomes, Peroxisomes, Vacuoles, and Vesicles

Vesicles and vacuoles are general terms for membranous sacs that store or carry substances. More specifically, **lysosomes** of animal cells contain digestive enzymes that damage the cell if they are released from their packaging into the cytosol. They are useful in self-destruction of old, damaged, or diseased cells. **Peroxisomes** are vesicles that contain oxidase and catalase, enzymes that degrade poisonous metabolic wastes such as free radicals and hydrogen peroxide. They are found in all types of eukaryotic cells but are prominent in the liver and kidney cells of mammals.

Mitochondria

Mitochondria are spherical to elongated structures with two phospholipid bilayer membranes found in most eukaryotes. Often called the "powerhouses" of the cell, they produce most of the ATP in many eukarytoic cells. The innermost membrane of a mitochondrion is folded into numerous **cristae** that increase the surface area for ATP production. Mitochondria contain 70S ribosomes and a circular molecule of DNA; however, most mitochondrial proteins are coded by nuclear DNA and synthesized by cytoplasmic ribosomes.

Chloroplasts

Chloroplasts are light-harvesting structures found in photosynthetic eukaryotes. Their pigments gather light energy to produce ATP and form sugar from carbon dioxide. Numerous membranous sacs called **thylakoids** form an extensive surface area for their biochemical and photochemical reactions (Chapter 5). Like mitochondria, chloroplasts have two phospholipid bilayer membranes, DNA, and ribosomes.

Endosymbiotic Theory

The **endosymbiotic theory** has been suggested to explain why mitochondria and chloroplasts have 70S ribosomes, circular DNA, and two membranes. The theory states that the ancestors of these organelles were prokaryotic cells that were internalized by other prokaryotes and then lost the ability to exist outside of their host—thus forming early eukaryotes. The theory is not universally accepted because it does not explain all of the facts.

KEY THEMES

Function is derived from structure, both at a molecular level and at the cellular level. Changes to structure ultimately lead to changes in function and affect the overall survival of microorganisms. As you read and study this chapter, it is important to form for yourself a firm mental image of what microbial cells look like. Specifically, you should focus on:

Prokaryotic microbes are fundamentally different from eukaryotic microbes: Though structurally less complex, prokaryotes are nonetheless arguably the most successful organisms on Earth. Their simplicity, however, does place a greater burden of survival on their ability to function. Knowing the differences between microbial structures is key to understanding microbial metabolism, genetic potential, and most aspects of their relationships with us.

QUESTIONS FOR FURTHER REVIEW

Answers to these questions can be found in the answer section at the back of this study guide. Refer to the answers only after you have attempted to solve the questions on your own.

Multiple Choice

1. Prokaryotes and eukaryotes display all of the common features of living organisms, but viruses do not. Of the characteristics listed below, the one that is seen inside a host cell is
   a. growth.
   b. reproduction.
   c. responsiveness.
   d. metabolism.
2. The key difference between prokaryotes and eukaryotes is that prokaryotes
   a. lack a cytoplasmic membrane.
   b. lack a nucleus.
   c. lack ribosomes.
   d. are always smaller than eukaryotes.
3. Which of the following are external structures found only in archaea?
   a. hami
   b. flagella
   c. pili
   d. fimbriae
4. Removing the glycocalyx from a prokaryotic cell could result in the cell
   a. drying out.
   b. becoming unable to attach to surfaces.
   c. being recognized by the immune system.
   d. undergoing all of the above.
5. A bacterial cell with flagella that cover the surface of the cell is called
   a. amphitrichous.
   b. polar.
   c. lophotrichous.
   d. peritrichous.
6. Which of the following structures in bacteria are not used for sticking to surfaces or other cells?
   a. fimbriae
   b. flagella
   c. pili
   d. All are used for attachment.
7. Of the characteristics listed below, which is true of both fimbriae and the cell wall?
   a. Both allow attachment to other cells.
   b. Both offer protection from immune recognition.
   c. Both allow for motility.
   d. Both offer protection from antimicrobial drugs.
8. Peptidoglycan is found in the cell walls of
   a. archaea.
   b. bacteria.
   c. fungi.
   d. algae.

9. Which of the following is NOT a function of the proteins found on the cytoplasmic membrane of any given cell?
   a. transport
   b. recognition
   c. macromolecular synthesis
   d. receptors
10. Of the following functions of the cytoplasmic membrane, which is (are) found in prokaryotes but NOT in eukaryotes?
    a. selective permeability
    b. energy storage
    c. group translocation
    d. both b and c
11. Based on the electrical gradient that forms across the cytoplasmic membrane, what types of molecules would be attracted to the inside of the cell?
    a. positively charged molecules
    b. negatively charged molecules
    c. neutral molecules
    d. both positively and negatively charged molecules
12. In passive transport mechanisms, energy is provided by
    a. ATP.
    b. concentration gradients.
    c. electrochemical gradients.
    d. Energy is not required in any form.
13. Which of the following could diffuse through the cytoplasmic membrane?
    a. a protein
    b. oxygen
    c. a phospholipid
    d. glucose
14. Group translocation is found in
    a. archaea.
    b. all bacteria.
    c. some bacteria.
    d. eukaryotes.
15. Inclusions within bacteria are used primarily for
    a. containing the nuclear material.
    b. containing excess nutrient materials.
    c. containing ribosomes for protein synthesis.
    d. inclusions form only in eukaryotes.
16. Which of the following microbes can possess a glycocalyx?
    a. bacteria
    b. protozoa
    c. viruses
    d. both a and b
17. Which of the following is a true statement regarding eukaryotic flagella?
    a. They are composed of flagellin.
    b. They are constructed exactly the same way as prokaryotic flagella.
    c. They are composed of tubulin.
    d. They move in a manner similar to the way in which prokaryotic flagella move.
18. Which of the following transport mechanisms occurs in eukaryotes but NOT in prokaryotes?
    a. facilitated diffusion
    b. active transport with symporters
    c. active transport with antiporters
    d. endocytosis
19. In eukaryotes, the closest structure to the prokaryotic nucleoid is
    a. the nuclear envelope.
    b. the nucleoplasm.
    c. the nucleolus.
    d. the chromatin.

20. The rough endoplasmic reticulum is rough because it contains
    a. lipids.
    b. proteins.
    (c) ribosomes.
    d. vesicles.

Fill in the Blanks

1. If the glycocalyx of a bacterium is well organized and firmly attached to the cell, it is called a capsule. Loosely constructed glycocalyces are called slime layer.
2. Bacterial flagella are composed of proteins called flagellin, whereas eukaryotic flagella are composed of tubulin.
3. The glycan portion of peptidoglycan is composed of alternating units of NAG and NAM.
4. The toxic part of a Gram-negative cell wall corresponds to lipid A which, along with sugar, forms the larger molecule lipid NAS.
5. The sterol-like molecules used by bacteria to stabilize the cytoplasmic membrane are called hopanoids.
6. For each scenario below, indicate the direction of movement for the molecules specified (use: "moves into the cell," "moves out of the cell," "does not move" to fill in the blanks).
   a. The concentration of sodium outside a cell is 25 μm, and the concentration of sodium inside the same cell is 2 μm. With respect to the cell, sodium moves inside the cell
   b. The concentration of potassium outside the cell is 10 μm, and the concentration of potassium inside the same cell is 10 μm. With respect to the cell, potassium does not move.
   c. Outside the cell, the concentration of sodium chloride is 20 μm, and the concentration of potassium chloride is 10 μm. Inside the cell, the concentration of sodium chloride is 30 μm, and the concentration of potassium chloride is 2 μm. If these compounds move as compounds, sodium chloride moves outside, and potassium chloride moves inside.

If these compounds move as ions, sodium ___out___, potassium ___in___, and chlorine ___out___.

7. Solutions across a cytoplasmic membrane with the same concentration of solutes and water are said to be ___isotonic___. If a hypertonic solution is outside the cytoplasmic membrane, water will move ___out of___ (into/out of) the cell, causing the cell to ___shrink___ (shrink/burst).
8. Symporters move two substances in ___the same___ (the same/different) direction(s) across a membrane. Antiporters move two substrates in ___different___ (the same/different) direction(s). Both of these transporters are examples of ___active___ (active/passive) transport.
9. Prokaryotic ribosomes are ___smaller___ (larger/smaller) than eukaryotic ribosomes. Overall, the prokaryotic ribosome is ___70___ S. The individual subunits are ___50___ S and ___30___ S.
10. The materials that make up the cell walls of eukaryotes are not the same materials that make up prokaryotic cell walls. Instead, plant cell walls are made of ___cellulose___, fungal cell walls are made of ___chitin___ and/or ___glucomannan___, and algal cell walls are composed of many different chemicals.
11. Membrane fluidity in eukaryotes is facilitated by sterols such as ___cholesterol___.
12. The cytoskeleton of eukaryotic cells is composed of ___~~phospolipid~~ microfilaments___, ___microtubules___, and ___intermediate___. (Give the names of the fibers and tubules, not the proteins from which they are made.)
13. The Golgi body packages molecules into ___proteins___ that travel to the cytoplasmic membrane, fuse, and release their contents by the process of ________________.

14. Within eukaryotic cells, the control center is the nu________________,

    and the powerhouse is the ________________.

15. The endosymbiotic theory is used to explain the origins of two eukaryotic

    organelles: ________________ and ________________.

Matching

Match the structure in the numbered list with the correct function or description in the alphabetical list. Answers will be used only once.

1. __F__ Glycocalyx
2. __J__ LPS
3. __H__ Cytoplasmic membrane
4. __I__ Lysosome
5. __B__ Mitochondria
6. __G__ Golgi body
7. __C__ Vacuoles
8. __A__ Flagella
9. __E__ Nucleoid
10. __D__ Fimbriae

A. Movement involving taxis
B. Generate energy in the form of ATP for the cell
C. Sacs used for cytoplasmic storage
D. Bristlelike appendages on the surface of prokaryotic cells
E. Area of the cytosol where DNA can be found
F. Protection and attachment for prokaryotic and eukaryotic cells
G. Involved in secretion of molecules to the outside of the cell by exocytosis
H. Selectively permeable barrier designed to regulate movement of molecules into and out of the cell
I. Breaks down nutrients, aids in the disposal of cellular material
J. Part of the outer membrane of Gram-negative bacteria; the lipid A portion is toxic

Short-Answer Questions for Thought and Review

1. Figures 3.2 and 3.3 show a comparison between prokaryotic and eukaryotic cells. Group the terms in the figures according to function, with each functional category containing both prokaryotic and eukaryotic counterparts.

2. Explain the process of positive taxis in which a bacterium moves toward a nutrient source.

3. Why might the corkscrew motility of spirochetes such as *Borrelia burgdorferi* aid in tissue invasion? What causes this motility?

4. Any internal bacterial infection is bad, but why would infections with Gram-positive bacteria be less damaging, in general, than those with Gram-negative bacteria? Your answer should relate specifically to the structure of the cell walls in these groups of organisms.

After cell death, Gram-neg release toxins

5. Draw a diagram of a cytoplasmic membrane showing an arrangement of active and/or passive transport mechanisms involving H^+ and glucose with the ultimate goal of getting glucose into the cell.

H+ H+ H+ H+ H+ H+ More outside?

6. When antimicrobial drugs are used, why is it best to use drugs that are specific to either microbial structure or function?

resistance

7. Why can't Gram-positive bacteria bring molecules into the cell by phagocytosis?

rigid cell wall

Critical Thinking

1. Of the general characteristics of living things—growth, reproduction, responsiveness, and metabolism—which one do you think is most important to the determination of whether something is alive or not? Why?
2. Eukaryotes are generally larger than prokaryotes, though there are exceptions (see Highlight: Giant Bacteria). Why might the presence of organelles inside eukaryotes allow them to consistently achieve a larger, more complex form?
3. During group translocation, glucose is converted to glucose 6-phosphate as it travels across the cytoplasmic membrane and enters the cell. What does the concentration gradient of glucose look like in this situation? In which direction will glucose flow—into the cell or out of the cell? Why?
4. Mitochondria and chloroplasts retain some DNA. If both organelles originated from engulfed prokaryotes, what types of gene remnants might you look for to help support the idea that they were once free-living? Refer to the general characteristics used to define living organisms to help you answer this question.

Concept Building Questions

1. In Chapter 2 we learned about various types of chemical bonds that form between atoms, molecules, compounds, and macromolecules. What types of bonds hold the bacterial cytoplasmic membrane together? In extremely hot environments, why would the single layer of branched lipids in archaea be more stable than the traditional lipid bilayer found in other cells? Answer in terms of chemical bonding.
2. How could Leeuwenhoek, with his primitive microscopes, have been able to discern differences between prokaryotic and eukaryotic cells? (Based on the structures described in this chapter, which of them could he have seen that would have allowed him to make the distinction?)

4 Microscopy, Staining, and Classification

CHAPTER SUMMARY

Units of Measurement (p. 97)

Scientists use metric units of measurement that are simpler than English units and are standardized throughout the world. The metric system is a decimal system in which each unit is one-tenth the size of the next largest unit. The basic unit of length in the metric system is the **meter (m)**, which is slightly longer than an English yard. One-thousandth of a meter is a millimeter (mm), about the thickness of a dime, and one thousandth of a millimeter is a micrometer (μm), which is small enough to be useful in measuring the size of cells. One-thousandth of a micrometer is a nanometer (nm), which is one-billionth of a meter and is used to measure the smallest cellular organelles and viruses.

Microscopy (pp. 98–107)

Microscopy refers to the use of light or electrons to magnify objects.

General Principles of Microscopy

The same general principles guide both light and electron microscopy.

Wavelength of Radiation

Various forms of radiation differ in **wavelength**, which is the distance between two corresponding parts of a wave. The human eye distinguishes different wavelengths of light as different colors. Moving electrons also act as waves, and their wavelengths depend on the voltage of an electron beam. Electron wavelengths are much smaller than those of visible light, and thus their use results in enhanced microscopy.

Magnification

Magnification is the apparent increase in size of an object and is indicated by a number followed by an "×," which is read "times." Magnification results when a beam of radiation refracts (bends) as it passes through a lens.

Resolution

Resolution (also called *resolving power*) is the ability to distinguish objects that are close together. The better the resolution, the better the ability to distinguish two objects. Modern microscopes can distinguish objects as close together as 0.2 μm. A principle of microscopy is that resolution distance depends on (1) the wavelength of the electromagnetic radiation and (2) the **numerical aperture** of the lens, which is its ability to gather light. **Immersion oil** is used to fill the space between the specimen and a lens to reduce light refraction and thus increase the numerical aperture and resolution.

Contrast

Contrast refers to differences in intensity between two objects, or between an object and its background. Because most microorganisms are colorless, they are stained to increase contrast. Polarized light may also be used to enhance contrast.

Light Microscopy

Several classes of microscopes use various types of light to examine specimens.

Bright-Field Microscopes

The most common microscopes are bright-field microscopes, in which the background (or field) is illuminated. There are two basic types: **simple microscopes**, which contain a single magnifying lens and are similar to a magnifying glass, and **compound microscopes**, which use a series of lenses for magnification. Light rays pass through a specimen and into an objective lens immediately above the object being magnified. The objective lens is really a series of lenses, and several objective lenses are mounted on a **revolving nosepiece**. The lenses closest to the eyes are **ocular lenses**, whereas **condenser lenses** lie beneath the stage of the microscope and direct light through the slide. The **total magnification** of a compound microscope is determined by multiplying the magnification of the objective lens by that of the ocular lens. The limit of useful magnification for light microscopes is 2000× because they are restricted by the wavelength of visible light. A photograph of a microscopic image is a **micrograph**.

Dark-Field Microscopes

Pale objects are best observed with **dark-field microscopes**, which utilize a dark-field stop in the condenser that prevents light from directly entering the objective lens. Instead, light passes into the slide at an oblique angle. Only light rays scattered by the specimen enter the objective lens and are seen, so the specimen appears light against a dark background.

Phase Microscopes

Phase microscopes use a **phase plate** to retard light rays passing through the specimen so that they are 1/2 wavelength out of phase with neighboring light waves, thereby producing contrast. **Phase-contrast microscopes** produce sharply defined images in which fine structures can be seen. **Differential interference contrast microscopes** create phase interference patterns and use prisms to split light beams into their component colors, giving images a dramatic three-dimensional or shadowed appearance.

Fluorescent Microscopes

Fluorescent microscopes use an ultraviolet (UV) light source to fluoresce objects. Because UV light has a shorter wavelength than visible light, resolution is increased. Contrast is improved because fluorescing structures are visible against a black background.

Confocal Microscopes

Confocal microscopes use fluorescent dyes in conjunction with UV lasers to illuminate the fluorescent chemicals in only a single plane of a specimen at a time. Several images are taken and digitized, and then computers construct three-dimensional images of the entire specimen.

Electron Microscopy

Because the shortest wavelength of visible light is about 400 nm, structures closer together than about 200 nm cannot be distinguished using light microscopy. By contrast, electrons traveling as waves have wavelengths between 0.01 nm and 0.001 nm; thus, their resolving power is much greater, and they typically magnify objects 10,000× to 100,000×. There are two general types.

Transmission Electron Microscopes

A **transmission electron microscope (TEM)** generates a beam of electrons that passes through a thinly sliced, dehydrated specimen, through magnetic fields that manipulate and focus the beam, and then onto a fluorescent screen that changes the electrons' energy into visible light.

Scanning Electron Microscopes

In a **scanning electron microscope (SEM)**, the surface of the specimen is first coated with a metal such as platinum or gold. The SEM then focuses the beam of electrons back and forth across the surface of the coated specimen, scanning it rather than penetrating it. Electrons scattered off the surface of the specimen pass through a detector and a photomultiplier, producing a signal that is displayed on a monitor.

Probe Microscopy

Probe microscopes use miniscule electronic probes to magnify specimens more than 100,000,000×. There are two types. **Scanning tunneling microscopes** pass a metallic probe across and above the surface of a specimen and measure the amount of electron flow. They can reveal details on a specimen surface at the atomic level. Atomic force microscopes traverse the tip of the probe lightly on the surface of the specimen. Deflection of a laser beam aimed at the probe's tip measures vertical movements that are translated by computer to reveal the specimen's atomic topography.

Staining (pp. 107–113)

Both light and electron microscopy use **staining**—the coloring of specimens with dyes—to increase contrast.

Preparing Specimens for Staining

Preparing specimens for staining involves making a thin film of organisms—or **smear**—on a slide, and then either passing the slide through a flame (**heat fixation**) or applying a chemical (**chemical fixation**) to attach the specimen firmly to the slide.

Principles of Staining

The colored portion of a dye, known as the **chromophore**, binds to chemicals via covalent, ionic, or hydrogen bonds. Anionic chromophores called **acidic dyes** or cationic chromophores known as **basic dyes** are used to stain different portions of an organism to aid viewing and identification.

Simple Stains

Simple stains are composed of a single basic dye such as crystal violet and involve no more than soaking the smear in the dye and rinsing.

Differential Stains

Differential stains use more than one dye so that different cells, chemicals, or structures can be distinguished. The **Gram stain** differentiates purple-staining Gram-positive cells from pink-staining Gram-negative cells. The procedure has four steps:

1. Flood the smear with the **primary stain**, crystal violet, and rinse.
2. Flood the smear with the **mordant**, iodine, and rinse.
3. Flood the smear with the **decolorizing agent**, a solution of ethanol and acetone, and rinse.
4. Flood the smear with the **counterstain**, safranin, and rinse.

The **acid-fast stain** is used to differentiate cells with waxy cell walls, such as cells of *Mycobacterium* and *Nocardia*. Endospores cannot be stained by normal techniques because their walls are practically impermeable to all chemicals. The **Schaeffer–Fulton endospore stain** uses heat to drive the primary stain, malachite green, into the endospore.

Histological Stains

Laboratory technicians commonly use the **Gomori methenamine silver** (**GMS**) and hematoxylin and eosin (HE) stains to identify features of histological specimens (tissue samples).

Special Stains

Acidic dyes are repulsed by the negative charges on the surface of cells and therefore do not stain them. Such dyes that stain the background and leave the cells colorless are called **negative (or capsule) stains. Flagellar stains** bind to flagella, increase their diameter, and change their color, all of which increases contrast and makes them visible.

Staining for Electron Microscopy

Stains used for TEM are chemicals containing atoms of heavy metals, such as lead, which absorb electrons.

Classification and Identification of Microorganisms (pp. 114–121)

Scientists sort organisms on the basis of mutual similarities into nonoverlapping groups called **taxa**. As discussed in Chapter 1, **taxonomy** is the science of classification.

Linnaeus and Taxonomic Categories

Carolus Linnaeus invented a system of taxonomy, grouping similar interbreeding organisms into **species**, species into **genera**, genera into **families**, families into **orders**, orders into **classes**, classes into **phyla**, and phyla into **kingdoms**. He gave each species a descriptive name consisting of a **genus name** and **specific epithet**. This practice of naming organisms with two names is called **binomial nomenclature**. Today, taxonomists place less emphasis on comparisons of physical and chemical traits, and greater emphasis on comparisons of genetic materials.

Domains

Carl Woese proposed the existence of three taxonomic **domains** based on three cell types revealed by rRNA sequencing: **Eukarya, Bacteria, and Archaea**. Cells of the three domains differ in many other characteristics, including their cell membrane lipids, transfer RNA (tRNA) molecules, and sensitivity to antibiotics.

Taxonomic and Identifying Characteristics

Taxonomists use one or more of five procedures to identify and classify microorganisms:

1. Many physical characteristics are used to identify microorganisms. For example, scientists can usually identify protozoa, fungi, algae, and parasitic worms based solely on their morphology. The appearance of bacterial colonies also gives clues to help identify microorganisms.
2. Microbiologists also use biochemical tests, noting a particular microbe's ability to utilize or produce certain chemicals.
3. Serological tests using **antiserum** (serum containing antibodies) can determine whether a microorganism produces an antigen-antibody reaction in the laboratory. In an **agglutination test**, antiserum is mixed with a sample that may be antigenic: clumping of antigen with antibodies (agglutination) indicates the presence of the target cells.
4. **Bacteriophages** (or simply **phages**) are viruses that infect and usually destroy bacterial cells. Whenever a specific phage is able to infect and kill bacteria, the resulting lack of bacterial growth produces a clear area within the bacterial lawn called a **plaque. Phage typing** may thus reveal that one bacterial strain is susceptible to a particular phage, whereas another is not, and allow scientists to distinguish between them.
5. Scientists also analyze a specimen's nucleic acids.

Taxonomic Keys

Microbiologists use **dichotomous keys**, which involve stepwise choices between paired characteristics, to assist in identifying microorganisms.

KEY THEMES

To study microbes, we need special tools to see them and ways in which to categorize them based on relationships of form and function. Microscopy and staining techniques are two of the primary methods by which we do both, though new methods of study are always being devised. As you study this chapter, focus on the following:

> *Microscopy allows us to magnify and distinguish microbial structures:* Different types of microscopes are used to look at microbes in different ways, but all exist to enhance our ability to see microorganisms. Staining techniques have also evolved over the years to aid viewing. Microscopes, therefore, provide a window onto the microbial world.
>
> *Classification allows microbiologists to study relationships:* By grouping microbes according to form, function, environmental relationships, and other parameters, we begin to see how microbes work together in the environment to sustain our world. These relationships will be explored throughout the rest of the textbook.

QUESTIONS FOR FURTHER REVIEW

Answers to these questions can be found in the answer section at the back of this study guide. Refer to the answers only after you have attempted to solve the questions on your own.

Multiple Choice

1. Magnification makes an object
   a. larger.
   b. appear larger.
   c. remain the same size.
   d. both b and c.
2. Which of the following could have been the most precisely resolved object seen under a Leeuwenhoek microscope?
   a. mitochondrion
   b. protein
   c. red blood cell
   d. virus
3. Why do most microbes have poor contrast?
   a. They are small.
   b. They are colorless.
   c. They don't let light pass through them.
   d. All of the above.
4. Oil immersion lenses help microscopists increase
   a. contrast.
   b. magnification.
   c. resolution.
   d. both b and c.
5. For a light microscope, the limit of resolution is determined by
   a. the quality of the lenses.
   b. the magnification of the lenses.
   c. contrast.
   d. the wavelength of visible light.
6. Living microbes can best be viewed using which type of microscopy?
   a. bright-field microscopy
   b. dark-field microscopy
   c. phase-contrast microscopy
   d. scanning electron microscopy
7. The resolving limit of the best light microscope under the best of conditions is __________, whereas that of a scanning electron microscope is __________.
   a. 200 μm, 10 μm
   b. 200 nm, 10 nm
   c. 200 μm, 0.001 nm
   d. 200 nm, 0.001 μm

8. What is the major difference between SEM and TEM?
   a. SEM doesn't view specimens under a vacuum.
   b. SEM requires specimens to be coated with metal "stains."
   c. SEM produces a three-dimensional image.
   d. SEM allows visualization of living organisms.

9. If you were performing a Gram stain on a bacterial sample and forgot the step that uses ethanol and acetone, what color would the cells be at the end of the Gram staining procedure?
   a. purple
   b. pink
   c. pink and purple, depending on cell wall type
   d. colorless

10. The most appropriate stain to use to identify a bacterial species that produces thick, waxy cell walls is the
    a. Gram stain.
    b. acid-fast stain.
    c. endospore stain.
    d. negative stain.

11. In staining, the purpose of a mordant is to
    a. serve as a primary stain.
    b. serve as a secondary stain or counterstain.
    c. decolorize cells.
    d. bind dyes to make them less soluble.

12. The major difference between the stains used in electron microscopy and those used in light microscopy is that stains for the electron microscope are
    a. chromophores.
    b. heavy metals.
    c. electron dense.
    d. both b and c.

13. The goal of modern classification schemes has shifted from a simple desire to catalog organisms to a desire to
    a. understand phylogenetic relationships.
    b. determine common ancestors.
    c. determine evolutionary relationships.
    d. All of the above are goals of modern taxonomy.

14. In the domain system of nomenclature, differences between organisms are based on
    a. rRNA sequences.
    b. mRNA sequences.
    c. DNA sequences.
    d. protein sequences.

15. Why are viruses NOT included among the three domains or five kingdoms of modern taxonomy?
    a. They are acellular.
    b. They are too small to be readily studied.
    c. They were not included in older taxonomic systems so should not be included in new systems.
    d. None of the above applies to viruses.

Fill in the Blanks

1. Perform the following calculations:

   a. Convert 5 nm to mm ____________

   b. Convert 2.5 dm to µm ____________

   c. Multiply 1 mm by 6 µm, and express your answer in nm ____________

2. Empty magnification produces images that are faint and blurry because the image lacks sufficient ____________________ and ____________________.

3. Fluorescent microscopes have ____________________ (better/worse) resolution than standard bright-field microscopes because UV light has a ____________________ (longer/shorter) wavelength than visible light.

4. The two types of microscopy that produce direct three-dimensional images for viewing are ____________________ and ____________________.

5. Electron microscopy is most useful for viewing a cell's ____________________, those parts of the cell that can't be seen any other way.

6. Staining is useful because it increases ____________________ between structures or a specimen and its background; this, in turn, allows for better ____________________.

7. Anionic chromophores are called ____________________ dyes and bind to ____________________ (acidic/basic) structures, whereas cationic chromophores are called ____________________ dyes and bind to ____________________ (acidic/basic) structures.

8. In order for cells exposed to a negative stain to be stained themselves, a ____________________ must be added, otherwise the cells remain colorless.

9. Taxonomy consists of three parameters: ____________________, ____________________, and ____________________.

10. In ____________________ tests, antiserum is mixed with patient samples. Antiserum contains ____________________ that recognize microbial ____________________.

Short-Answer Questions for Thought and Review

1. Bioterrorism is, unfortunately, a real threat to our future. One agent that could be used is anthrax (Bacillus anthracis). Based on what you have read

about microscopy and staining in this chapter, propose a relatively simple method for identifying anthrax quickly.

2. Discuss how the introduction of artifacts into microscopic specimens during preparation for viewing could prevent a laboratory technician from making a proper identification of a microbe.

3. Compare and contrast the purposes for using a simple stain versus using a differential stain. Why are differential stains more useful?

4. The five kingdoms once widely accepted (Animalia, Plantae, Fungi, Protista, and Prokaryotae) are now contained in three domains. What are the three domains, and which kingdoms are in which domains?

5. Rank the commonly used techniques for identifying microbes—serology, biochemistry, morphology, and nucleic acid analysis—in order of most specific to least specific in terms of accuracy in identification.

Critical Thinking

1. Objective lenses on a compound light microscope come in several magnifications, usually 4×, 10×, 40×, and 100× (oil immersion). When you first put a slide on the microscope stage to locate your specimen, which lens should you use, and why?
2. Some methods of classification and identification are much more specific than others (e.g., nucleic acid analysis versus Gram stain). Medical diagnostics rely on proper identification of infectious agents for proper treatment. Why, then, are most diagnostic tools composed of methods that are generally less precise?

Below is a dichotomous key. Use the information in the key to answer the questions that follow.

1a Gram-positive cells	2
1b Gram-negative cells	Gram-negative bacteria
2a Rods	3
2b Cocci	(omitted from chart for simplicity)
3a Catalase positive	4
3b Catalase negative	7
4a Produces endospores	5
4b Does not produce endospores	6
5a Acid from mannitol	non-*cereus Bacillus* species
5b Does not produce acid from mannitol	*Bacillus cereus*
6a Acid from glucose	*Corynebacterium xerosis*
6b Does not produce acid from glucose	*Corynebacterium pseudodiphtheriticum*
7a Gas from glucose	*Lactobacillus fermentum*
7b Does not produce gas from glucose	8
8a Acid from mannitol	*Lactobacillus casei*
8b Does not produce acid from manitol	*Lactobacillus acidophilus*

3. While working as a medical technologist, you culture a bacterium from a patient sample that has the following characteristics: Gram-positive rod that is catalase positive, does not produce endospores, and does not produce acid from glucose. According to the key, what organism did you culture?
4. Take a closer look at the key. Notice that both step 5 and step 8 involve determining acid formation from mannitol. What does this tell you about the ability to properly identify an organism? (Hint: Think about how much information you can get from any one test.)

Concept Building Questions

1. In Chapter 3 we learned about the differences between prokaryotic and eukaryotic structures. Many of these structures are in fact very different between the two groups, but not all. Morphology was one of the early classification schemes used to group organisms into taxonomic schemes.
   a. Use the tick lineage depicted in Figure 4.22 on page 116 of the text to explain why a binomial nomenclature system based on physical characteristics could lead to mistakes in classifying and identifying organisms.
   b. Go back to Table 3.6 on page 88 that compares prokaryotes (archaea and bacteria) and eukaryotes. Which of the characteristics, based on morphology, would be enough to distinguish prokaryotes from eukaryotes? Which ones would not?
   c. Could any of the characteristics ascribed specifically to bacteria be used to differentiate species based solely on appearance? Why or why not?

5 Microbial Metabolism

CHAPTER SUMMARY

Basic Chemical Reactions Underlying Metabolism (pp. 127–135)

Catabolism and Anabolism

Metabolism is the sum of complex biochemical reactions within an organism. **Catabolic reactions** break down nutrient molecules and generate ATP molecules (exergonic), and **anabolic reactions** synthesize macromolecules and use ATP energy (endergonic).

Enzymes catabolize nutrients into **precursor metabolites,** which are rearranged by polymerization reactions to form macromolecules. Cells grow as they assemble these large molecules into cell parts. Reproduction usually occurs when the cell has doubled in size.

Oxidation and Reduction Reactions

Oxidation–reduction (redox) reactions involve the transfer of electrons. These reactions always occur simultaneously because an electron donated by one chemical is accepted by another chemical. The electron acceptor is said to be reduced. The molecule that loses an electron is **oxidized.** If the electron is part of a hydrogen atom, the reaction is called a **dehydrogenation reaction.**

Three electron carrier molecules that are often required in metabolic pathways are **nicotinamide adenine dinucleotide** (NAD^+), nicotinamide adenine dinucleotide phosphate ($NADP^+$), and **flavin adenine dinucleotide (FAD).**

ATP Production and Energy Storage

Energy from the chemical bonds of nutrients is concentrated in the high-energy phosphate bonds of ATP. **Substrate-level phosphorylation** describes the transfer of phosphate from a phosphorylated organic nutrient to ADP to form ATP. **Oxidative phosphorylation** phosphorylates ADP using inorganic phosphate and energy from respiration. **Photophosphorylation** is the phosphorylation of ADP with inorganic phosphate using energy from light. There is a cyclical conversion of ATP from ADP and back with the gain and loss of phosphate.

The Roles of Enzymes in Metabolism

Catalysts increase reaction rates of chemical reactions but are not permanently changed in the process. **Enzymes,** organic catalysts, are often named for their **substrates,** which are the chemicals they cause to react. Substrates fit onto the specifically shaped **active sites** of enzymes.

Naming and Classifying Enzymes

Enzymes are classified into six categories based on their mode of action: **hydrolases** add hydrogen and hydroxide from the hydrolysis of water to split larger molecules into smaller ones; **isomerases** rearrange atoms within a molecule; **ligases** or polymerases join molecules; **lyases** split molecules without using water; **oxidoreductases** oxidize or reduce; and **transferases** transfer functional groups.

The Makeup of Enzymes

Many protein enzymes are complete in themselves. Other enzymes are composed of **apoenzymes**—a protein portion—and one or more nonprotein cofactors. Inorganic cofactors include ions such as iron, magnesium, zinc, or copper. Organic cofactors are also called **coenzymes.** Organic cofactors are made from vitamins and include NAD^+, $NADP^+$, and FAD. The combination of an apoenzyme and its cofactors is a **holoenzyme**.

RNA molecules functioning as enzymes are called **ribozymes.** Ribozymes process RNA molecules in eukaryotes. Ribosomal enzymes catalyze the actual protein synthesis reactions of ribosomes; thus, ribozymes make protein enzymes.

Enzyme Activity

Activation energy is the amount of energy required to initiate a chemical reaction. Activation energy may be supplied by heat, but high temperatures are not compatible with life; therefore, enzymes are required to lower the activation energy needed. The complementary shapes of active sites of enzymes and their substrates determine enzyme-substrate specificity. In catabolism, an enzyme binds to a substrate, forming an enzyme-substrate complex; the bonds within the substrate are broken; the enzyme separates from the two new products; and the enzyme is released to act again.

Enzymes may be **denatured** by physical and chemical factors such as temperature and pH, which change their shape and thus their ability to bond. The change may be reversible or permanent.

The rate of enzymatic activity is also affected by the concentrations of substrate and enzyme. Enzyme activity proceeds at a rate proportional to the concentration of substrate molecules until all the active sites on the enzymes are filled to saturation.

Enzyme activity may be blocked by **competitive inhibitors**, which block but do not denature active sites. In **allosteric inhibition, noncompetitive inhibitors** attach to an allosteric site on an enzyme, distorting the active site and halting enzymatic activity. In excitatory allosteric control, the change in the shape of the active site activates an inactive enzyme.

Feedback inhibition (**negative feedback**) occurs when the final product of a series of reactions is an allosteric inhibitor of some previous step in the series. Thus, accumulation of the end product "feeds back" a stop signal to the process.

Carbohydrate Catabolism (pp. 135–148)

Glycolysis

Glycolysis (the **Embden-Meyerhof pathway**) involves the splitting of a glucose molecule in a series of 10 steps that ultimately results in two molecules of pyruvic acid and a net gain of two ATP and two NADH molecules. The 10 steps of glycolysis can be divided into three stages: energy investment, lysis, and energy conserving.

Cellular Respiration

Cellular respiration is a three-stage metabolic process that involves oxidation of substrate molecules and production of ATP. The stages of respiration are synthesis of acetyl-CoA, the Krebs cycle, and electron transport.

Synthesis of Acetyl-CoA and the Krebs Cycle

Acetyl-coenzyme A (**acetyl-CoA**) is formed when two carbons from pyruvic acid join coenzyme A. Two molecules of acetyl-CoA, two molecules of CO_2, and two molecules of NADH are produced. Acetyl-CoA enters the **Krebs cycle**, a series of enzymatic steps that transfers energy and electrons from acetyl-CoA to coenzymes NAD^+ and FAD. For every two molecules of acetyl-CoA that enter the Krebs cycle, two molecules of ATP, six molecules of NADH, and two molecules of $FADH_2$ are formed.

Electron Transport

The **electron transport chain** is a series of redox reactions that passes electrons from one membrane-bound carrier to another and then to a **final electron acceptor**. The energy from these electrons is used to pump protons (H^+) across the membrane. Ultimately, ATP is synthesized.

The four categories of carrier molecules in the electron transport system are flavoproteins, ubiquinones, proteins containing heavy metal, and cytochromes.

Aerobes use oxygen atoms as final electron acceptors in the electron transport chain in a process known as **aerobic respiration**, whereas anaerobes use other inorganic molecules such as sulfate, nitrate, and carbonate as final electron acceptors in **anaerobic respiration**.

Chemiosmosis

Chemiosmosis is a mechanism in which the flow of ions down an **electrochemical gradient** across a membrane is used to synthesize ATP. For example, energy released during the redox reactions of electron transport is used to pump protons across a membrane, creating a proton gradient.

A **proton gradient** is an electrochemical gradient of protons that has potential energy known as a proton motive force. When protons flow down their electrochemical gradient through protein channels called ATPases, ATP is synthesized. **ATP synthases (ATPases)** are enzymes that synthesize ATP by oxidative phosphorylation and photophosphorylation.

About 34 ATP molecules are synthesized per pair of electrons traveling an electron transport chain. Thus, there is a theoretical net yield of 38 ATP molecules from the aerobic respiration of 1 molecule of glucose via glycolysis (4 molecules of ATP produced minus 2 molecules of ATP used), the Krebs cycle (2 molecules of ATP produced), and the electron transport chain (34 molecules of ATP produced).

Alternatives to Glycolysis

The **pentose phosphate** and **Entner-Doudoroff pathways** are alternative pathways for the catabolism of glucose, but they yield fewer ATP molecules than does the Embden-Meyerhof pathway. However, they produce precursor metabolites not produced in glycolysis. The pentose phosphate pathway produces metabolites used in synthesis of nucleotides, amino acids, and glucose by photosynthesis.

The Entner-Doudoroff pathway, used by only a few bacteria, uses different enzymes and yields precursor metabolites and NADPH, which is not produced by the Embden-Meyerhof or pentose phosphate pathways.

Fermentation

Fermentation is the partial oxidation of sugar to release energy using an organic molecule within the cell as the final electron acceptor. In lactic acid fermentation, NADH reduces pyruvic acid from glycolysis to form lactic acid. In alcohol fermentation, pyruvic acid undergoes decarboxylation (CO_2 is given off) and reduction by NADH to form ethanol. Some fermentation products are useful to health and industry, and some are harmful.

Other Catabolic Pathways (pp. 148–149)

Lipid Catabolism

Fats are catabolized by lipases that break the glycerol–fatty acid bonds via hydrolysis. Glycerol is converted to DHAP to be catabolized by glycolysis and the Krebs cycle. Fatty acids are catabolized by **beta-oxidation reactions** that form acetyl-CoA and generate NADH and $FADH_2$.

Protein Catabolism

Protein catabolism by prokaryotes involves **protease** enzymes secreted to digest large proteins outside their cell walls. The resulting amino acids move into the cell and are used in anabolism or **deaminated** to produce substrates for the Krebs cycle.

Photosynthesis (pp. 149–154)

Chemicals and Structures

Photosynthesis is a process in which light energy is captured by pigment molecules (called **chlorophylls**) and transferred to ATP and metabolites. **Photosystems**, photosystem I (PS I) and photosystem II (PS II), are networks of chlorophyll molecules and other pigments held within a protein matrix in membranes called **thylakoids.**

Prokaryotic thylakoids are infoldings of the cytoplasmic membrane, whereas eukaryotic thylakoids are infoldings of the inner membranes of chloroplasts. Stacks of thylakoids within chloroplasts are called grana.

Light-Dependent Reactions

The light absorption and redox reactions of photosynthesis are classified as **light-dependent reactions** (light reactions) and **light-independent reactions** (dark reactions). The latter synthesize glucose from carbon dioxide and water regardless of light conditions.

A **reaction center chlorophyll** is a special chlorophyll molecule of photosystem I, which is excited by transferred energy absorbed by pigment molecules elsewhere in the photosystem. Excited electrons from the reaction center are passed to an acceptor of an electron transport chain, protons are pumped across the membrane, a proton motive force is created, and ATP is generated in a process called **photophosphorylation.**

Cyclic Photophosphorylation

In **cyclic photophosphorylation**, electrons return to the original reaction center chlorophyll after passing down the electron transport chain. The resulting proton gradient produces ATP by chemiosmosis.

Noncyclic Photophosphorylation

In **noncyclic photophosphorylation**, photosystem II works with photosystem I, and the electrons are used to reduce $NADP^+$ to NADPH. Therefore, in noncyclic photophosphorylation, a cell must constantly replenish electrons to PS II. In oxygenic organisms, the electrons come from H_2O. In anoxygenic organisms, the electrons come from inorganic compounds such as H_2S.

Light-Independent Reactions

ATP and NADPH from the light-dependent reactions drive the synthesis of glucose in the light-independent pathway of photosynthesis. The **Calvin-Benson cycle** of the light-independent pathway occurs in three steps: **carbon fixation**, in which CO_2 is reduced; reduction by NADPH to form molecules of G3P, which join to form glucose; and regeneration of RuBP to continue the cycle.

Other Anabolic Pathways (pp. 154–158)

Because anabolic reactions are synthesis reactions, they require energy and metabolites, both of which are often the products of catabolic reactions. **Amphibolic reactions** are metabolic reactions that can proceed toward catabolism or toward anabolism, depending on the needs of the cell. Examples are found in the biosynthesis of carbohydrates, lipids, amino acids, and nucleotides.

Carbohydrate Biosynthesis

Gluconeogenesis refers to metabolic pathways that produce sugars, starch, cellulose, glycogen, peptidoglycan, and so on, from noncarbohydrate precursors such as amino acids, glycerol, and fatty acids.

Lipid Biosynthesis

Lipids are synthesized by a variety of routes. Fat is synthesized from glycerol and three molecules of fatty acid—a reverse of the catabolic reaction. Steroids result from complex pathways involving polymerizations and isomerizations of sugar and amino acid metabolites.

Amino Acid Biosynthesis

Amino acids are synthesized by **amination**, a process in which the amine group from ammonia is added to a precursor metabolite, and by **transamination**, a reversible reaction in which an amine group is transferred from one amino acid to another by the action of enzymes using coenzyme pyridoxal phosphate.

Nucleotide Biosynthesis

Nucleotides are produced from precursor metabolites derived from glycolysis and the Krebs cycle: ribose and deoxyribose from ribose-5 phosphate, phosphate from ATP, and purines and pyrimidines from the amino acids glutamine and aspartic acid.

Integration and Regulation of Metabolic Functions (pp. 158–159)

Energy released in catabolic reactions is used to drive anabolic reactions. Catabolic pathways produce metabolites to use as substrates for anabolic reactions.

The pathways of cellular metabolism can be categorized into three groups: pathways synthesizing macromolecules (proteins, nucleic acids, polysaccharides, and lipids), intermediate pathways, and pathways that produce ATP and precursor molecules (glycolysis, the Krebs cycle, the pentose phosphate pathway, and the Entner-Doudoroff pathway).

Cells use a variety of mechanisms to regulate metabolism, including **control of gene expression**, which controls enzyme production needed for metabolic pathways, and **control of metabolic expression**, in which the cells control enzymes that have been produced.

KEY THEMES

So far we have looked at history, basic chemistry, microbial structure and function, and methodologies for studying microbes. One common thread running through all of these chapters is microbial metabolism. Microbiologists have been concerned with metabolic function since the early days of microbiology. Chemistry is the reason metabolism works, and much of the structure and function of the cell is given over to ensuring that metabolism continues. Many of the methods for identifying microbes depend on metabolic differences between organisms. As you study this chapter, keep this one fundamental concept in mind:

> *The survival of all life depends on metabolism:* Catabolism drives anabolism, which is the foundation for new life; energy is derived from and used by these processes. Life ends when metabolism stops, for microbes and for us. Knowing how microbes work helps us to understand how we work, and helps keep us alive.

QUESTIONS FOR FURTHER REVIEW

Answers to these questions can be found in the answer section at the back of this study guide. Refer to the answers only after you have attempted to solve the questions on your own.

Multiple Choice

1. The ultimate goal of metabolism is to allow the organism to
   a. grow.
   b. reproduce.
   c. respond to the environment.
   d. move about its environment.

2. Any given catabolic pathway can produce
   a. ATP only.
   b. metabolites only.
   c. both ATP and metabolites.
   d. all of the above.
3. In the reaction $NAD^+ \rightarrow NADH$, oxidation is occurring by
   a. gain of an electron.
   b. loss of a hydrogen atom.
   c. gain of a hydrogen atom.
   d. As written, the reaction shows reduction, not oxidation.
4. Which of the following involves production of energy from respiration?
   a. substrate-level phosphorylation
   b. oxidative phosphorylation
   c. photophosphorylation
   d. all of the above
5. The chemical reactions that sustain life depend on which of the following to keep them going?
   a. high concentrations of reactants
   b. high temperatures
   c. random collisions between reactants
   d. organic catalysts called enzymes
6. The movement of the phosphate from glucose 6-phosphate to ADP would be performed by what type of enzyme?
   a. isomerase
   b. transferase
   c. oxidoreductase
   d. hydrolase
7. Cofactors are
   a. nonprotein substances.
   b. small organic molecules.
   c. part of a holoenzyme.
   d. all of the above.
8. Which of the following methods is NOT used by a cell to regulate enzyme function?
   a. synthesis of enzymes only when needed
   b. sequestering of enzymes in organelles
   c. feedback inhibition
   d. All of the above are methods of regulating enzyme function.
9. In what way are competitive inhibitors and noncompetitive inhibitors similar?
   a. Both bind the active site of an enzyme.
   b. Both bind away from the active site of an enzyme.
   c. Both can turn enzymes off.
   d. Both can turn enzymes on.
10. The primary energy source that is oxidized to allow anabolism to occur is
    a. carbohydrates.
    b. lipids.
    c. proteins.
    d. nucleic acids.
11. Which of the following is common to both respiration and fermentation?
    a. electron transport chain
    b. Krebs cycle
    c. glycolysis
    d. beta-oxidation
12. Substrate-level phosphorylation can best be described as the direct transfer of phosphate
    a. from ATP to substrates.
    b. from ADP to substrates.
    c. between ADP and ATP molecules.
    d. from one substrate to another.

13. What is the end product of glycolysis?
   a. 2 phosphoenopyruvic acids
   b. 2 pyruvic acids
   c. 2 acetyl-CoAs
   d. 2 glyceraldehyde 3-phosphates

14. The pentose phosphate pathway is used primarily to
   a. serve as a backup in case the cell can no longer use glycolysis.
   b. produce more ATP than glycolysis when the cell needs more energy quickly.
   c. produce precursor metabolites used in anabolism.
   d. produce NADH, which is not made by glycolysis.

15. The pentose phosphate pathway provides what essential metabolite to photosynthesis?
   a. glucose 6-phosphate
   b. ribulose 5-phosphate
   c. glyceraldehyde 3-phosphate
   d. pyruvic acid

16. What is the final oxidation product of glucose following cellular respiration?
   a. pyruvic acid
   b. carbon dioxide
   c. acetyl-CoA
   d. ATP

17. Which system is responsible for producing the most ATP?
   a. glycolysis
   b. Krebs cycle
   c. electron transport chain
   d. All produce equal amounts of ATP.

18. Coenzyme Q is an example of which of the four categories of carrier molecules used in the electron transport chain?
   a. flavoproteins
   b. ubiquinones
   c. metaloproteins
   d. cytochromes

19. How does oxidative phosphorylation differ from substrate-level phosphorylation?
   a. It generates ATP, and substrate-level phosphorylation does not.
   b. It generates ATP by using a proton motive force.
   c. It is not used to generate ATP.
   d. It produces less ATP than substrate-level phosphorylation.

20. Which of the following could be a final electron acceptor in fermentation?
   a. oxygen
   b. sulfate
   c. organic molecules
   d. All of the above can be used.

21. The major purpose of fermentation is to
   a. produce ATP from ADP.
   b. produce NAD^+ from NADH.
   c. produce alcohol.
   d. take the place of glycolysis.

22. Which of the following is true of beta-oxidation?
   a. It creates acetyl-CoA by breaking down the hydrocarbon portion of lipids.
   b. It produces DHAP for glycolysis.
   c. It generates NADH for the electron transport chain.
   d. All of the above are true of beta-oxidation.

23. Bacteriochlorophylls are found in which photosynthetic organism?
   a. algae
   b. cyanobacteria
   c. green and purple bacteria
   d. archaea

24. During photosynthesis, glucose is synthesized from carbon dioxide and water using
    a. light-dependent reactions.
    b. light-independent reactions.
    c. the Calvin-Benson cycle.
    d. both b and c.

25. ATP is produced in photosynthesis by
    a. substrate-level phosphorylation.
    b. chemiosmosis.
    c. aerobic respiration.
    d. anaerobic respiration.

26. Which of the following pathways does NOT contribute precursor metabolites to the synthesis of amino acids?
    a. glycolysis
    b. Krebs cycle
    c. pentose phosphate pathway
    d. gluconeogenesis

27. The majority of metabolic pathways are
    a. amphibolic.
    b. anabolic.
    c. catabolic.
    d. Few pathways are amphibolic, but there are roughly equal numbers of catabolic and anabolic pathways.

Fill in the Blanks

1. Catabolism provides ______________ and ______________ to fuel anabolism. In general, catabolic pathways are ______________ (exergonic/endergonic), which means they ______________ (release/require) energy.

2. In redox reactions, electron ______________ give electrons to another molecule and are said to be ______________. The molecule that accepts the electron is an electron ______________ and is said to be ______________.

3. The three important electron carriers found in metabolic pathways are ______________, ______________, and ______________.

4. Enzyme inhibitors can either be ______________ or ______________. The ______________ type of inhibition can sometimes be overcome by simply adding more of the enzyme's substrate to the reaction.

5. Glucose is the favored "input" for catabolism. It is broken down (catabolized) either by respiration or by fermantation (name the general metabolic processes).

6. Two pathways can take the place of glycolysis, at least partially. These pathways are the pentose (P) and the ED. All three pathways produce pyruvate, which can be sent to the Krebs cycle during cellular respiration.

7. Indicate where the following pathways occur in prokaryotic and eukaryotic cells:

Pathway	*Prokaryote*	*Eukaryote*
Glycolysis		mito
Krebs cycle		cyto
Electron transport chain		
Photosynthesis		chloroplast

8. The electron transport chain moves electrons from acceptor to acceptor, causing the transfer of protons across the membrane to establish a proton gradient; ATP is then generated by the process of chemiosmosis.

9. The final electron acceptor in aerobic respiration is oxygen, whereas the final electron acceptor in anaerobic respiration is generally inorganic molecules

10. During the catabolism of proteins, ____________________ cleave the proteins into amino acids that have their amino groups removed by ____________________ prior to being funneled into the ____________________ to be recycled.

11. ____________________ photophosphorylation uses only

____________________, which serves as the initial electron donor and

final electron acceptor. ____________________ photophosphorylation

uses ____________________ and ____________________.

12. Glycolysis and gluconeogenesis are examples of ____________________

pathways.

Matching

Use the numbered items to match the pathway with the molecules generated. Each letter may be used more than once, and each molecule may have more than one answer.

1. _____ Glyceraldehyde 3-phosphate
2. _____ ATP
3. _____ NADH
4. _____ NAD^+
5. _____ NADPH
6. _____ Glucose
7. _____ CO_2
8. _____ Ribose 5-phosphate
9. _____ Pyruvic acid
10. _____ Glucose 6-phosphate

A. Glycolysis
B. Pentose phosphate pathway
C. Entner-Doudoroff pathway
D. Calvin-Benson cycle
E. Krebs cycle
F. Electron transport chain
G. Gluconeogenesis

Short-Answer Questions for Thought and Review

1. Explain the process of allosteric inhibition. How can this type of inhibition be overcome?

2. Fermentation produces much less energy than respiration, but it is used by many cells. Why is fermentation necessary for the cells that perform it?

3. Why must the rate of anabolism be linked to the rate of catabolism in the cell? For example, what would happen if catabolism proceeded at a much slower rate than anabolism?

4. Write a complete chemical equation for one of the four redox reactions that occur in the Krebs cycle. Indicate which molecules are electron donors, which are electron acceptors, which are oxidized, and which are reduced.

Critical Thinking

1. What advantage does a cell that can make everything it needs from precursor metabolites have over a cell that must acquire some of these metabolites from nutrient sources in the environment? What is the disadvantage? Why is it best to be able to do both?
2. Glycolysis and gluconeogenesis have opposite reactions but use many of the same enzymes. How does the cell keep the process it needs going in one direction in a situation like this? For example, how does the cell make glucose with the same pathway that wants to break it down? Relate your answer to the pathways themselves and to what you have learned about the regulation of enzymes.
3. Explain as specifically as you can what happens to ATP production when a bacterial cell growing in an oxygenated environment suddenly finds itself in an anaerobic environment.

Concept Building Questions

1. A microbe living in the environment will catabolize proteins, bring the parts into the cell, and, through anabolism, rebuild new proteins. Because many of the proteins used in the cell are the same across species, why can't microbes simply import whole proteins and not expend ATP to make them? Relate your answer to microbial structure, which was discussed in Chapter 3.
2. Enzymes reduce the activation energy needed to allow chemical reactions to occur by creating enzyme-substrate complexes that are conducive to the reaction. What types of bonds can form between an enzyme and a substrate that will help reduce activation energy but still allow the reaction to occur? What types of bonds can't form between an enzyme and a substrate? If these "wrong" bonds formed in the complex, what could happen to the reaction, and why?

6 Microbial Nutrition and Growth

CHAPTER SUMMARY

Growth Requirements (pp. 167–175)

Microbiologists use the term **growth** to indicate an increase in the size of a **population** of microbes rather than an increase in size of an individual. Microbial growth depends on the metabolism of nutrients and results in the formation of a discrete **colony**, an aggregation of cells arising from a single parent cell. A **nutrient** is any chemical required for growth of microbial populations. The most important of these are compounds containing carbon, oxygen, nitrogen, and/or hydrogen.

Nutrients: Chemical and Energy Requirements

All cells require three things to conduct metabolism: a carbon source, a source of energy, and a source of electrons or hydrogen atoms.

Sources of Carbon, Energy, and Electrons

Organisms can be categorized into one of four groups based on their source of carbon and their use of either chemicals or light as a source of energy:

1. **Photoautotrophs** use carbon dioxide as a carbon source and light energy from the environment to make their own food.
2. **Chemoautotrophs** use carbon dioxide as a carbon source but catabolize organic molecules for energy.
3. **Photoheterotrophs** are photosynthetic organisms that acquire energy from light and acquire nutrients via catabolism of organic compounds.
4. **Chemoheterotrophs** use organic compounds for both energy and carbon.

In addition, organotrophs acquire electrons from organic molecules that provide them carbon, whereas lithotrophs acquire electrons from inorganic molecules.

Oxygen Requirements

Obligate aerobes require oxygen as the final electron acceptor of the electron transport chain, whereas **obligate anaerobes** cannot tolerate oxygen and use an electron acceptor other than oxygen. Toxic forms of oxygen are highly reactive and cause a chain of vigorous oxidation.

Four forms of oxygen are toxic:

1. **Singlet oxygen** ($^1O_2$) is molecular oxygen with electrons that have been boosted to a higher energy state, typically during aerobic metabolism. Phototrophic microorganisms often contain pigments called carotenoids that prevent toxicity by removing the excess energy of singlet oxygen.

2. **Superoxide radicals** ($O_2^{\cdot -}$) are formed during the incomplete reduction of oxygen during electron transport in aerobes and during metabolism by anaerobes in the presence of oxygen. They are detoxified by superoxide dismutase.
3. **Peroxide anion** (O_2^{2-}) is a component of hydrogen peroxide, which is formed during reactions catalyzed by superoxide dismutase. The enzymes catalase and peroxidase detoxify peroxide anion.
4. **Hydroxyl radicals** ($OH^{\cdot}$) result from ionizing radiation and from the incomplete reduction of hydrogen peroxide. Hydroxyl radicals are the most reactive of the four toxic forms of oxygen, but because hydrogen peroxide does not accumulate in aerobic cells, the threat of hydroxyl radicals is virtually eliminated in aerobic cells.

Not all organisms are strictly aerobes or anaerobes. **Facultative anaerobes** can maintain life via fermentation or anaerobic respiration, though their metabolic efficiency is often reduced in the absence of oxygen. **Aerotolerant anaerobes** prefer anaerobic conditions but can tolerate oxygen because they have some form of the enzymes that detoxify oxygen's poisonous forms. **Microaerophiles** require low levels of oxygen.

Nitrogen Requirements

Nitrogen is a growth-limiting nutrient for many microorganisms, which acquire it from organic and inorganic nutrients. Though nitrogen constitutes about 79% of the atmosphere, relatively few organisms can utilize nitrogen gas. A few bacteria reduce nitrogen gas to ammonia via a process called **nitrogen fixation**, which is essential to life on Earth.

Other Chemical Requirements

In addition to the main elements found in microbes, very small amounts of **trace elements** such as selenium, zinc, and so on, are required. Most microorganisms also require small amounts of certain organic chemicals that they cannot synthesize. These are called **growth factors**. For example, vitamins are growth factors for some microorganisms.

Physical Requirements

In addition to chemical nutrients, organisms have physical requirements for growth, including specific conditions of temperature, pH, osmolarity, and pressure.

Temperature

Because both proteins and lipids are temperature sensitive, different temperatures have different effects on the survival and growth rates of microbes. Though microbes survive within the limits imposed by a minimum growth temperature and a maximum growth temperature, an organism's metabolic activities produce the highest growth rate at the **optimum growth temperature.**

Microbes are described in terms of their temperature requirements as (from coldest to warmest): **psychrophiles** require temperatures below 20°C; **mesophiles** grow best at temperatures ranging between 20°C and 40°C; **thermophiles** require temperatures above 45°C; and **hyperthermophiles** require temperatures above 80°C.

pH

Organisms are sensitive to changes in acidity because hydrogen ions and hydroxyl ions interfere with hydrogen bonding within proteins and nucleic acids; as a result, organisms have ranges of acidity that they prefer and can tolerate. Most bacteria and protozoa are called **neutrophiles** because they grow best in a narrow range around a neutral pH, between 6.5 and 7.5. By contrast, other bacteria and many fungi are **acidophiles**, and grow best in acidic environments where pH can range as low as 0.0. **Alkalinophiles** live in alkaline soils and water up to pH 11.5.

Physical Effects of Water

Microorganisms require water to dissolve enzymes and nutrients and to act as a reactant in many metabolic reactions. Osmotic pressure restricts cells to certain environments. Whereas the cell walls of some microbes protect them from osmotic shock, osmosis can cause other cells to die from either swelling and bursting or shriveling (**crenation**). **Obligate halophiles** require high osmotic pressure such as exists in saltwater. **Facultative halophiles** do not require but can tolerate salty conditions.

Water exerts pressure in proportion to its depth, and the pressure in deep ocean basins and trenches is tremendous. Organisms that live under extreme pressure are called **barophiles**. Their membranes and enzymes depend on pressure to maintain their three-dimensional functional shapes, and typically they cannot survive at sea level.

Associations and Biofilms

Relationships in which one organism harms or even kills another are considered **antagonistic**. In **synergistic** relationships, members of an association cooperate such that each receives benefits that exceed those that would result if each lived separately. In **symbiotic** relationships, organisms live in close nutritional or physical contact, becoming interdependent.

Biofilms are an example of complex relationships among numerous microorganisms, often of different species, that attach to surfaces through the production of an extracellular matrix and display metabolic and structural traits different from those expressed by the microorganisms living individually. They often form as a result of **quorum sensing**, a process in which microorganisms respond to the density of nearby microorganisms by utilizing signal and receptor molecules.

Culturing Microorganisms (pp. 175–183)

Microbiologists culture microorganisms by transferring an **inoculum**—a sample—from a clinical or environmental specimen into a **medium**, a collection of nutrients. Liquid media are called **broths**. Microorganisms that grow from an inoculum are called a **culture**. Cultures visible on the surface of solid media are called **colonies**.

Clinical Sampling

A **clinical specimen** is a sample of human material, such as feces, saliva, cerebrospinal fluid, or blood, which is examined and tested for the presence of microorganisms. Clinical specimens must be properly labeled and transported to a microbiological laboratory to avoid both death of the pathogens and growth of normal organisms.

Obtaining Pure Cultures

Suspected pathogens must be isolated from the normal microbiota in culture. Scientists use several techniques to isolate organisms in **pure cultures** (**axenic cultures**) composed of cells arising from a single progenitor called a **colony-forming unit** (**CFU**). To obtain pure cultures, all media, vessels, and instruments must be sterile; that is, free of any microbial contaminants. The use of aseptic techniques is critical as well.

The most commonly used isolation technique in microbiological laboratories is the **streak-plate method.** In this technique, a sterile inoculating loop is used to spread an inoculum across the surface of a solid medium in Petri dishes. After an appropriate period of time called **incubation,** colonies develop from each isolate and are distinguished from one another by differences in characteristics.

In the **pour-plate technique,** CFUs are separated from one another using a series of dilutions. The final dilutions are mixed with warm agar in Petri dishes. Individual CFUs form colonies in and on the agar.

Culture Media

A variety of media are available for microbiological cultures. A common example is **nutrient broth. Agar,** a complex polysaccharide, is a useful compound because it is difficult for microbes to digest, solidifies at temperatures below 40°C, and does not melt below 100°C. Still-warm liquid agar media can be poured into Petri dishes, to make **Petri plates.** When warm agar media are poured into test tubes that are then placed at an angle and left to cool until the agar solidifies, the result is **slant tubes,** or slants. Other culture media include the following:

A medium for which the precise chemical composition is known is called a **defined medium** (or **synthetic medium**).

Complex media contain a variety of growth factors and can support a wider variety of microorganisms than can defined media.

Selective media typically contain substances that either favor the growth of particular microorganisms or inhibit the growth of unwanted ones.

Differential media are formulated such that either the presence of visible changes in the medium or differences in the appearances of colonies helps microbiologists differentiate among the different kinds of bacteria growing on the medium. One example involves the differences in organisms' utilization of the red blood cells in blood agar.

Reducing media provide conditions conducive to culturing anaerobes. They contain compounds that chemically combine with free oxygen and remove it from the medium.

Transport media are used by health care personnel to move specimens safely from one location to another while maintaining the relative abundance of organisms and preventing contamination of the specimen or its environment.

Special Culture Techniques

Special culture techniques include the following:

Animal cultures and **cell cultures** allow for the growth of microorganisms for which artificial media are inadequate. Mammals, bird eggs, and cultures of living cells are used.

Low-oxygen cultures favor the growth of microorganisms that thrive in environments intermediate between strictly aerobic and anaerobic, such as within the respiratory or intestinal tract of mammals. Candle jars, chemical packets, or carbon dioxide incubators are used to remove oxygen from the environment.

Enrichment cultures use a selective medium designed to increase very small numbers of a chosen microbe to observable levels.

Cold-enrichment cultures require the incubation of a specimen in a refrigerator, allowing for the enrichment of the culture with cold-tolerant species.

Preserving Cultures

Refrigeration at 4°C is often the best technique for storing bacterial cultures for short periods of time. Deep-freezing and lyophilization are used for long-term storage of bacterial cultures. **Deep-freezing** involves freezing the cells at temperatures from –50°C to –95°C. **Lyophilization** is freeze-drying—that is, removal of water from a frozen culture via an intense vacuum.

Growth of Microbial Populations (pp. 183–191)

Most unicellular microorganisms reproduce by binary fission, a process in which a cell grows to twice its normal size and then divides in half to produce two daughter cells of equal size.

Mathematical Considerations in Population Growth

With binary fission, any given cell divides to form two cells; then each of these new cells divides in two, for a total of four cells, and then four become eight, and so on. This type of growth, called **logarithmic growth** or **exponential growth,** produces dramatically greater yields than simple addition, known as arithmetic growth. Microbiologists use scientific notation to deal with the huge numbers involved in expressing microbial population size.

Generation Time

The time required for a bacterial cell to grow and divide is its **generation time.** Viewed another way, generation time is the time required for a population of cells to double in number. Most bacteria have a generation time of 1–3 hours.

Phases of Microbial Population Growth

A graph that plots the number of bacteria growing in a population over time is called a **growth curve.** When microbial growth is plotted on a **semilogarithmic scale** (which uses a logarithmic scale for the y-axis), the plot of the population's growth results in a straight line. When bacteria are grown in a broth, the typical microbial growth curve has four distinct phases:

In the **lag phase,** the organisms are adjusting to their environment.

In the **log phase,** the population is most actively growing.

In the **stationary phase,** new organisms are being produced at the same rate at which they are dying.

In the **death phase,** the organisms are dying more quickly than they can be replaced by new organisms.

Continuous Culture in a Chemostat

A special culture device called a chemostat can be used to maintain a particular phase of microbial population growth. A **chemostat** is an open system in which fresh medium is added and old medium is removed maintaining steady nutrient levels.

Measuring Microbial Reproduction

Because of each cell's small size and incredible rate of reproduction, it is not possible to count every one in a population. Thus, microbiologists estimate population size by counting the number in a small, representative sample, and then multiplying. Microbiologists use either direct or indirect methods to estimate the number of cells.

Among the many direct methods are the following:

In **viable plate counts**, microbiologists base an estimate of the size of a microbial population on the number of colonies formed when diluted samples are plated onto agar media.

In **membrane filtration**, a large sample is poured through a filter small enough to trap cells.

The **most probable number (MPN) method** is a statistical estimation technique based on the fact that the more bacteria are in a sample, the more dilutions are required to reduce their number to zero.

In **microscopic counts**, a sample is placed on a cell counter, a glass slide with an etched grid, and viewed through a microscope. A microbiologist can count the number of bacteria in several of the large squares and then calculate the mean number of bacteria per square.

Electronic counters are devices that count cells as they interrupt an electrical current flowing across a narrow tube held in front of an electronic detector. Flow cytometry is one variation.

Among the indirect methods are measurements of metabolic activity, measurements of a population's dry weight, and measurement of the turbidity of a broth, especially by a device known as a **spectrophotometer**. Scientists can estimate numbers of unculturable microbes using genetic methods such as DNA analysis.

KEY THEMES

In the last chapter we learned about metabolism, an intricate assortment of chemical pathways essential to life. In Chapter 3 we learned how microbes bring in nutrients from the environment to fuel metabolism. In this chapter we focus on those nutrients from two perspectives: what microbes need in their natural environment to sustain themselves, and what materials must be present in the laboratory to mimic nature. While studying this chapter, focus on the following:

A microbe must either obtain the nutrients it needs from the environment or be able to produce everything it needs itself: Though each microbe is different, all must obtain the materials essential for survival; if they do not, they die.

Growth of populations proceeds through phases: Nutrients are not constant in the environment; how microbes grow will therefore vary depending on many parameters.

QUESTIONS FOR FURTHER REVIEW

Answers to these questions can be found in the answer section at the back of this study guide. Refer to the answers only after you have attempted to solve the questions on your own.

Multiple Choice

1. An organism that uses glucose for carbon and sulfur compounds for energy is a
   a. photoautotroph.
   b. photoheterotroph.
   c. chemoautotroph.
   d. chemoheterotroph.
2. Which of the groups of microbes below is most metabolically diverse in terms of nutritional classes?
   a. algae
   b. prokaryotes
   c. fungi
   d. protozoa
3. Which of the oxygen species below is NOT toxic to cells?
   a. O_2
   b. $O_2^{\cdot-}$
   c. O_2^{2-}
   d. All are toxic to cells.
4. Hydroxyl radicals are neutralized in the cell by the action of
   a. carotenoids.
   b. superoxide dismutase.
   c. peroxidase.
   d. All of the above work to neutralize hydroxyl radicals.
5. Which element listed below is often growth limiting in the environment?
   a. carbon
   b. hydrogen
   c. nitrogen
   d. oxygen
6. Vitamins are examples of
   a. trace elements.
   b. growth factors.
   c. antioxidants.
   d. both b and c.
7. A microbe's optimum growth temperature is generally closer to
   a. the minimum growth temperature.
   b. the maximum growth temperature.
   c. the midpoint between the minimum and maximum growth temperatures.
   d. The optimum temperature is the same as the maximum temperature.
8. An organism living at the bottom of the ocean would have to be
   a. barophilic.
   b. psychrophilic.
   c. anaerobic.
   d. all of the above.
9. Biofilms are best described as
   a. antagonistic relationships between microbes.
   b. synergistic relationships between microbes.
   c. symbiotic relationships between microbes.
   d. complex relationships showing features of synergy and symbiosis.
10. Of the methods used to isolate CFUs, which one would allow you to isolate microaerophilic microbes?
    a. streak plate
    b. pour plate
    c. streak plate incubated in a candle jar
    d. both b and c

11. On what type of media would you try to grow an organism whose nutritional requirements were not known?
    a. defined media
    b. complex media
    c. anaerobic media
    d. selective media
12. On what type of media would you try to grow and differentiate two species of digestive tract microorganisms?
    a. defined media
    b. complex media
    c. anaerobic media
    d. selective media
13. What type of special culture technique would need to be employed to propagate viruses in the laboratory?
    a. cell culture
    b. low-oxygen culture
    c. enrichment culture
    d. None of the above could be used to propagate a virus.
14. To preserve a bacterial specimen for many years, it should be stored
    a. at room temperature.
    b. in a refrigerator.
    c. in a deep freezer.
    d. as a lyophilized sample.
15. If a population of bacteria is growing in a sewer line with a continuous flow of wastes, which phase of the growth cycle will the population remain in most of the time?
    a. lag phase
    b. log phase
    c. stationary phase
    d. death phase
16. In which phase of the growth curve are antimicrobial treatments most successful?
    a. lag phase
    b. log phase
    c. stationary phase
    d. death phase
17. A person with an active infection begins to take an antibiotic. The drug should push the bacterial population into which phase of the growth curve?
    a. lag phase
    b. log phase
    c. stationary phase
    d. death phase
18. Which of the following techniques would best determine the number of living cells in a culture?
    a. viable plate count
    b. microscopic count
    c. Coulter counter
    d. most probable number

Fill in the Blanks

1. Organisms must acquire ____________________ from the environment to provide energy and building materials for anabolism.
2. Facultative anaerobes ____________________ (do/do not) have to have mechanisms to protect against the effects of toxic oxygen species. An example of a facultative anaerobe is ____________________.

3. Nitrogen gas in the atmosphere is turned into useful biological forms by prokaryotes performing this process: ____________________.

4. ____________________ grow best at temperatures lower than 15°C; thermophiles, in contrast, grow best above ____________________°C.

5. Fungi are generally ____________________ (acidophiles/neutrophiles/alkalinophiles).

6. Microbes can be grown in liquid ____________________ cultures or as ____________________ on solid agar plates.

7. The most common isolation technique used in the microbiology laboratory is the ____________________ method.

8. ____________________ cultures are designed to increase the population of a desired microbe that is present in a sample in very small numbers.

9. A bacterial culture starts with five cells. The bacteria reproduce every 30 minutes. At the end of 6 hours, you would have ____________________ cells in your culture.

10. ____________________ is a(n) ____________________ (direct/indirect) method for estimating the growth of a microbial population in broth that uses spectrophotometry.

Short-Answer Questions for Thought and Review

1. What is a limiting nutrient? Is hydrogen a good example of a limiting nutrient? Why?

2. Explain why aerotolerant and microaerophilic microbes can't survive high oxygen levels but can survive low oxygen levels.

3. In regard to the environmental requirements of microbes (pH, oxygen, etc.), the terms “tolerant” and “obligate” are often used. Give one example of why being tolerant is better than being obligate.

4. How could the routine ingestion of over-the-counter antacids contribute to the formation of ulcers by Helicobacter pylori?

5. Clinical samples from patients must be handled so as to avoid contamination. What are some of the ways a clinical sample can be contaminated from the time it is taken from the patient to the time it is analyzed in the laboratory?

Critical Thinking

1. There is a very large difference between the amount of carbon comprising a prokaryotic cell (50%) and the amount of iron (0.25%). Both elements are necessary to the cell, and the cell would die without either. Why, then, is there such a disparity in how much of each is needed? (Hint: When composing your answer, think of what the elements are used for.)
2. Which type of microbe, a single-celled microbe or a multicellular microbe, will have a broader temperature range over which it can survive? Why?
3. Where might you expect to find a fastidious microorganism living?

Concept Building Questions

1. Explain how limiting nutrients in the environment affects the growth potential of microbes. Relate your answer to transport mechanisms across the cytoplasmic membrane (Chapter 3) and to metabolism (Chapter 5).
2. When white blood cells are isolated from blood products, it is sometimes desired to remove the red blood cells. If the blood cells have been resuspended in a saline solution (0.9% NaCl), should the solution added to lyse the red blood cells have a higher or lower concentration of NaCl? Why?
3. Figure 6.8 presents some of the characteristics of bacterial colonies. These descriptions are often useful when trying to determine the identity of a bacterial sample. In Chapter 4 we learned that morphological characteristics alone generally are not enough to identify the bacteria with certainty. Explain why the information in Figure 6.8 can help narrow the possible identities of a microbe but can’t be used to make a definitive identification.

7 Microbial Genetics

CHAPTER SUMMARY

The Structure and Replication of Genomes (pp. 197–205)

Genetics is the study of inheritance and inheritable traits. **Genes** are composed of specific sequences of nucleotides that code for polypeptides or RNA molecules. A **genome** is the sum of all the genetic material in a cell or virus. Prokaryotic and eukaryotic cells use DNA as their genetic material; some viruses use DNA, and other viruses use RNA.

The Structure of Nucleic Acids

The two strands of DNA are held together by hydrogen bonds between complementary bases of nucleic acids called **base pairs (bp)**. In DNA, adenine bonds with thymine, and guanine bonds with cytosine. One end of a DNA strand is called the 5' end because it terminates in a phosphate group attached to a 5' carbon; the opposite end of the strand is called the 3' end because it terminates with a hydroxyl group bound to a 3' carbon of deoxyribose. The two strands are oriented in opposite directions to each other: one strand runs in a 3' to 5' direction, whereas the other runs in a 5' to 3' direction. The lengths of DNA molecules are expressed in base pairs.

The Structure of Prokaryotic Genomes

Prokaryotic genomes consist of one or two **chromosomes**, which are typically circular molecules of DNA associated with protein and RNA molecules, localized in a region of the cytoplasm called the **nucleoid**. Prokaryotic cells may also contain one or more extrachromosomal DNA molecules called **plasmids**, which contain genes that regulate nonessential life functions such as bacterial conjugation; resistance to one or more antimicrobial drugs, heavy metals, or toxins; destruction of competing bacteria; and pathogenicity.

The Structure of Eukaryotic Genomes

In addition to DNA, eukaryotic chromosomes contain proteins called **histones**, arranged as **nucleosomes** (beads of DNA) that clump with other proteins to form **chromatin fibers**. Eukaryotic cells also contain extrachromosomal DNA in mitochondria, chloroplasts, and plasmids.

DNA Replication

DNA replication is a simple concept: A cell separates the two original strands and uses each as a template for the synthesis of a new complementary strand. The process is **semiconservative** because each daughter DNA molecule is composed of one original strand and one new strand.

DNA replication begins at a specific nucleotide sequence called an **origin**. DNA helicase unzips the double helix, breaking hydrogen bonds between complementary base pairs, to form a replication fork. DNA synthesis always moves in the 5' to 3' direction, so the **leading strand** is synthesized toward the replication fork. Synthesis is mediated by enzymes that prime, join, and proofread the pairing of new nucleotides. The **lagging strand** is synthesized in a direction away from the replication fork, and discontinuously in Okazaki fragments. It always lags behind the process occurring in the leading strand. DNA ligase seals the gaps between adjacent Okazaki fragments to form a continuous DNA strand.

After bacterial DNA replication, methylation occurs. In **methylation**, a cell adds a methyl group to one or two bases that are part of specific nucleotide sequences. In some cases, genes that are methylated are "turned off" and are not transcribed, whereas in other cases, they are "turned on" and are transcribed. In some bacteria, methylated nucleotide sequences play a role in initiating DNA replication, repairing DNA, or recognizing and protecting against viral DNA.

Eukaryotic DNA replication is similar to that in bacteria, with a few exceptions. Eukaryotic cells use four DNA polymerases to replicate DNA. Because of the large size of eukaryotic chromosomes, there are many origins of replication. Okazaki fragments of eukaryotes are smaller than those of bacteria. Finally, plant and animal cells methylate cytosine bases exclusively.

Gene Function (pp. 206–220)

To understand gene function, it is necessary to distinguish between an organism's genotype from its phenotype.

The Relationship between Genotype and Phenotype

The **genotype** of an organism is the actual set of genes in its genome, whereas the **phenotype** is the physical and functional traits expressed by those genes, such as the presence of flagella. Thus, genotype determines phenotype; however, not all genes are active at all times.

The Transfer of Genetic Information

The **central dogma** of genetics states that DNA is transcribed to RNA, which is translated to form polypeptides.

The Events in Transcription

The transfer of genetic information begins with **transcription** of the genetic code from DNA to RNA, in which **RNA polymerase** links RNA nucleotides that are complementary to genetic sequences in DNA. Transcription begins at a region of DNA called a **promoter** (recognized by RNA polymerase) and ends with a sequence called a **terminator**. Other proteins may assist in termination, or it may depend solely on the nucleotide sequence of the transcribed RNA.

Cells transcribe four types of RNA from DNA:

RNA primer molecules for DNA polymerase to use during DNA replication.

Messenger RNA (mRNA) molecules, which carry genetic information from chromosomes to ribosomes.

Ribosomal RNA (rRNA) molecules, which combine with ribosomal polypeptides to form ribosomes, the organelles that synthesize polypeptides.

Transfer RNA (tRNA) molecules, which deliver amino acids to the ribosomes.

Eukaryotic transcription differs from bacterial transcription in several ways. Eukaryotic cells transcribe RNA in the nucleus, whereas prokaryotic transcription occurs in the cytosol. Eukaryotes have three types of nuclear RNA polymerase and multiple transcription factors. Eukaryotic cells process mRNA before translation. RNA processing involves capping, polyadenylation, and splicing.

Translation

In **translation**, the sequence of genetic information carried by mRNA is used by ribosomes to construct polypeptides with specific amino acid sequences. To understand how four DNA nucleotides can specify the 21 different amino acids commonly found in proteins requires an understanding of the genetic code. Scientists define the genetic code as the complete set of triplets of mRNA nucleotides called **codons** that code for specific amino acids. These bind to complementary **anticodons** on tRNA. The code is redundant; that is, more than one codon is associated with all the amino acids except methionine and tryptophan.

The smaller subunit of a ribosome is shaped to accommodate three codons at one time. Each ribosome also has three binding sites that are named for their function:

The **A site** accommodates a tRNA delivering an amino acid.

The **P site** holds a tRNA and the growing polypeptide.

Discharged tRNAs exit from the **E site.**

Prokaryotic translation proceeds in three stages: In **initiation**, an initiation complex is formed. During **elongation**, tRNAs sequentially deliver amino acids as directed by the codons of mRNA. Ribosomal RNA in the large ribosomal subunit catalyzes a peptide bond between the amino acid at the A site and the growing polypeptide at the P site. The third stage, termination, does not involve tRNA; instead, proteins called **release factors** halt elongation. The ribosome then dissociates into its subunits.

Regulation of Genetic Expression

Many genes are expressed at all times; other genes are regulated so that the polypeptides they encode are synthesized only when a cell has need of them. Cells may regulate synthesis by initiating or stopping transcription or by stopping translation directly.

Cells can control which mRNA molecules are translated into polypeptides. A **riboswitch** is a molecule of mRNA that folds in such a way as to favor or block translation of the polypeptide they encode depending on the need for that polypeptide. Translation can also be controlled by **small interfering RNA (siRNA)**. siRNA is an RNA molecule complementary to a portion of mRNA, tRNA, or a gene. siRNA binds to its target and renders it inactive. **Micro RNAs (miRNAs)** function similar to siRNA but only bind complementary mRNA.

An **operon** consists of a promoter, an adjacent regulatory element called an **operator**, and a series of genes whose transcription is determined by a regulatory gene located outside the operon. **Inducible operons** such as the **lac operon** are not usually transcribed and must be activated by inducers. **Repressible operons** such as the trp operon are transcribed continually until deactivated by repressors.

Mutations of Genes (pp. 220–227)

A mutation is a change in the nucleotide sequence of a genome.

Types of Mutations

Mutations range from large changes in an organism's genome, such as the loss or gain of an entire chromosome, to the most common type of mutation, **point mutations**, in which just one or a few nucleotide base pairs are affected. Point mutations include the following:

Substitutions, in which a single nucleotide is substituted for another, possibly leaving the amino acid sequence unaffected because of the redundancy of the genetic code.

Frameshift mutations, including **insertions** and **deletions** of nucleotides, in which nucleotide triplets following an insertion or deletion are displaced, creating new sequences of codons that result in vastly altered polypeptide sequences.

Effects of Mutations

Some base-pair substitutions produce **silent mutations**. Because of the redundancy of the genetic code, the substitution does not change the amino acid sequence. A change in a nucleotide sequence that specifies a different amino acid is called a **missense mutation**; what gets transcribed and translated makes sense, but not the right sense. In a **nonsense mutation**, a base-pair substitution changes an amino acid codon into a stop codon. Nearly all nonsense mutations result in nonfunctional proteins. Frameshift mutations (insertions or deletions) typically result in drastic missense and nonsense mutations.

Mutagens

Mutations can be spontaneous or result from recombination. Physical or chemical agents called **mutagens**, which include radiation and several types of DNA-altering chemicals, induce mutations. Radiation in the form of X rays and gamma rays can cause mutations. Additionally, nonionizing radiation in the form of ultraviolet light causes adjacent pyrimidine bases to bond to one another to form **pyrimidine dimers**. The presence of dimers prevents hydrogen bonding with the nucleotides in the complementary strand, distorts the sugar-phosphate backbone, and prevents proper replication and transcription.

Chemical mutagens include **nucleotide analogs**, compounds that are structurally similar to normal nucleotides but, when incorporated into DNA, cause mutations. Some **nucleotide-altering chemicals** alter the structure of nucleotides, causing base-pair substitution mutations. Aflatoxins are nucleotide-altering chemicals that result in missense mutations and cancer. Still other mutagenic chemicals insert or delete nucleotide base pairs, resulting in frameshift mutations.

Frequency of Mutation

About 1 of every 10 million genes contains an error. Mutagens typically increase the mutation rate by a factor of 10–1000 times.

DNA Repair

Cells have numerous methods of repairing damaged DNA. In **light repair**, cells use DNA photolyase to break the bonds between adjoining pyrimidine nucleotides. In **dark repair**, enzymes repair pyrimidine dimers by cutting damaged DNA from the molecule, creating a gap that is repaired by DNA polymerase I and DNA ligase. In **base-excision repair**, an enzyme system excises the erroneous base, and DNA polymerase I fills in the gap. In **mismatch repair**, enzymes scan newly synthesized unmethylated DNA looking for mismatched bases, which they remove and replace. Once a new DNA strand is methylated, mismatch repair enzymes cannot correct any errors that remain. When damage is so extensive that these mechanisms are overwhelmed, bacterial cells resort to an **SOS response** involving the production of a novel DNA polymerase capable of copying less-than-perfect DNA.

Identifying Mutants, Mutagens, and Carcinogens

If a cell does not repair a mutation, it and its descendants are called **mutants**. In contrast, cells normally found in nature are called **wild-type cells**. Researchers have developed methods to recognize mutants amidst their wild-type neighbors. These include:

- **Positive selection**, which involves selecting a mutant by eliminating wild-type phenotypes.
- **Negative selection** (also called **indirect selection**), a process in which a researcher attempts to culture auxotrophs.
- The **Ames test**, which is used to identify potential **carcinogens** (cancer-causing agents).

Genetic Recombination and Transfer (pp. 227–233)

Genetic recombination refers to the exchange of nucleotide sequences between two DNA molecules, often mediated by segments that are composed of identical or nearly identical nucleotide sequences called **homologous sequences**. DNA molecules that contain new arrangements of nucleotide sequences are called **recombinants**.

Horizontal Gene Transfer among Prokaryotes

Vertical gene transfer is the transmission of genes from parents to offspring. In **horizontal (lateral) gene transfer**, DNA from a **donor cell** is transmitted to a **recipient cell**. A **recombinant cell** results from genetic recombination between donated and recipient DNA.

Transformation, transduction, and bacterial conjugation are types of horizontal gene transfer:

- In **transformation**, a **competent** recipient cell takes up DNA from the environment. Competency is found naturally in only a few bacteria and can be created artificially in other bacteria.
- In **transduction**, a virus such as a bacteriophage carries DNA from a donor cell to a recipient cell. Donor DNA is accidentally incorporated in **transducing phages**.
- In **conjugation**, a bacterium containing an **F (fertility) plasmid**, or **F factor**, forms a **conjugation pilus** that attaches to an F^- recipient bacterium. Plasmid

genes are transferred to the recipient, which becomes F^+ as a result. **Hfr (high frequency of recombination) cells** result when an F plasmid integrates into a prokaryotic chromosome. Hfr cells form conjugation pili and transfer cellular genes more frequently than normal F^+ cells.

Transposons and Transposition

Transposons are DNA segments that code for the enzyme transposase and contain palindromic sequences known as **inverted repeats (IR)** at each end. (A palindrome is a word, phrase, or sentence that has the same sequence of letters when read backward or forward.) Transposons move among locations in chromosomes in eukaryotes and prokaryotes. The simplest transposons, known as **insertion sequences (IS)**, consist only of inverted repeats and transposase. **Complex transposons** contain other genes as well.

KEY THEMES

So far we have focused on the structural elements of microbial cells and the processes of metabolism and growth. We have yet to address the origin of these structures and functions. Evolution ultimately works on the phenotype, but it is the genotype, the sum total of a specie's genes, that ultimately produces the visible phenotypes. While studying this chapter on genetics, focus on the following key points:

Genomes are large, even among microbes, but not all of what they code for is expressed: Only a small fraction of the entire genome is actually translated into protein. Expression is highly regulated, a fact essential to survival.

Mutation is the foundation of evolutionary adaptation: Without the process of mutation, genes would not change, and ultimately there would be no variation among organisms.

Many microbes exchange genes horizontally: Horizontal gene transfer allows for the rapid spread of genetic elements among "adult" populations. This can be enormously beneficial to the microbes, but devastating for us.

QUESTIONS FOR FURTHER REVIEW

Answers to these questions can be found in the answer section at the back of this study guide. Refer to the answers only after you have attempted to solve the questions on your own.

Multiple Choice

1. DNA genomes are found in
   a. all organisms and all viruses.
   b. all organisms, but only some viruses.
   c. eukaryotic cells only.
   d. all eukaryotic cells, but only some prokaryotic cells and some viruses.

2. Plasmid DNA
   a. is the same as genomic DNA.
   b. is DNA in addition to genomic DNA.
   c. generally codes for extra functions.
   d. both b and c

3. Which of the following cells does NOT use histones to compact DNA?
   a. archaea
   b. bacteria
   c. eukaryotic cells
   d. All use histones to compact DNA.
4. Plasmids CANNOT be found among representatives of which of the following microbes?
   a. algae
   b. archaea
   c. bacteria
   d. fungi
5. Leading strand synthesis is initiated by
   a. a DNA primer.
   b. an RNA primer.
   c. an amino acid primer.
   d. a protein primer.
6. Which of the following molecules is needed in lagging strand synthesis but NOT in leading strand synthesis?
   a. primer
   b. polymerase
   c. ligase
   d. helicase
7. Which of the following is NOT a role of methylation?
   a. initiating DNA replication
   b. regulating gene expression
   c. identifying host DNA from viral DNA
   d. All are functions of methylation.
8. Of the types of RNA listed below, which is(are) NOT translated into protein?
   a. mRNA
   b. rRNA
   c. tRNA
   d. both b and c
9. Transcription in prokaryotes requires which of the following molecules that is NOT used in eukaryotic transcription?
   a. RNA polymerase
   b. sigma factor
   c. transcription factors
   d. All are needed in eukaryotic transcription.
10. Which of the following statements is true concerning DNA polymerases and RNA polymerases?
    a. Both require the aid of a helicase.
    b. RNA polymerase is faster than DNA polymerase.
    c. RNA polymerase uses ribonucleotides, whereas DNA polymerase uses deoxyribonucleotides.
    d. Both have proofreading capabilities.
11. In prokaryotes, which codon serves the dual purpose of being a start signal for translation and coding for N-formylmethionine?
    a. UAG
    b. UGA
    c. AUG
    d. GUA
12. Which of the following is true regarding eukaryotic mRNA?
    a. It always codes for only a single polypeptide.
    b. Translation of the mRNA can begin before transcription is complete.
    c. mRNA is transcribed in the cytosol.
    d. None of the above is true about eukaryotic mRNA.
13. Anticodons are found as part of
    a. mRNA structure.
    b. tRNA structure.
    c. rRNA structure.
    d. ribosomal structure.

14. Which stage of translation does NOT require energy?
    a. initiation
    b. elongation
    c. termination
    d. All steps require energy.

15. Which of the following is NOT part of the initiation complex in prokaryotic translation?
    a. mRNA
    b. $tRNA^{MET}$
    c. small ribosomal subunit
    d. All are part of the initiation complex.

16. Termination of translation requires
    a. tRNA.
    b. protein release factors.
    c. recognition of a stop codon.
    d. both b and c.

17. Which of the following is a true statement concerning the lac operon in E. coli?
    a. The lac operon is an inducible operon that is generally inactive.
    b. Lactose is an inducer of the operon.
    c. The lac operon allows for the conservation of energy by making catabolic enzymes for lactose only when lactose is present.
    d. All of the above are true statements.

18. Which type of mutation almost always results in a completely nonfunctional protein?
    a. silent mutation
    b. missense mutation
    c. frameshift mutation
    d. nonsense mutation

19. Pyrimidine dimers are caused by
    a. ionizing radiation.
    b. ultraviolet light.
    c. nucleotide analogs.
    d. nucleotide-altering chemicals.

20. Horizontal gene transfer among microbes
    a. occurs at a high frequency in microbial populations.
    b. involves donor and recipient cells within the same generation.
    c. does not include the method of transformation.
    d. does not occur at all.

21. Which of the following is a true statement regarding transduction?
    a. Transduction involves replicating viruses that move nonviral DNA between cells.
    b. Transduction occurs only among prokaryotic cells.
    c. Transduction involves both generalized and specialized processes.
    d. Both a and c are true.

22. Donor cells remain alive in which method of horizontal gene transfer?
    a. transformation
    b. transduction
    c. conjugation
    d. Donor cells die in all three processes.

23. Which process below is most likely to create frameshift mutations in the DNA of cells?
    a. transformation
    b. transduction
    c. conjugation
    d. transposition

24. Hfr conjugation
    a. creates F^+ cells out of the recipient.
    b. requires competent recipient cells.
    c. involves transfer of part of the donor chromosome.
    d. requires transposase.

25. Which of the molecules or entities listed below could not only carry an antibiotic resistance gene but also move the gene into a recipient cell with a high degree of success?
    a. complex transposon
    b. insertion sequence
    c. Hfr cell
    d. transducing phage

Fill in the Blanks

1. In RNA, __uracil__ replaces __thymine__ to bond with adenine.
2. Eukaryotic genomes, as compared with prokaryotic genomes, are usually __linear__ (linear/circular). Additionally, eukaryotic genomes generally have __more__ (more/fewer) chromosomes than prokaryotic genomes.
3. Monomers and energy needed to anabolically synthesize new strands of DNA are provided by the same molecule, specifically __tri-(P) DNA__.
4. During semiconservative replication, the __leading__ strand is synthesized continuously, whereas the __lagging__ strand is synthesized in short pieces.
5. The process of __transcription__ produces RNA templates from DNA, whereas __translation__ produces protein from RNA templates.
6. Transcription can terminate in one of two ways: __enzyme-dep.__ or __self__.
7. In mRNA, nucleotides called __codons__ code for specific amino acids.
8. The process of transcription produces three types of RNA: __m__, __t__, and __r__. These RNA molecules are then used in the process of __translation__.
9. Transfer RNA is __specific__ (specific/nonspecific) for the amino acids carried. Individual tRNA molecules can recognize __more than one__ (one/more than one) codon.

10. In translation, the ___A___ site receives incoming amino acids, the ___P___ site is involved in elongation, and the ___E___ site releases spent tRNAs.

11. Operons consist of the following elements: ___promoter___, ___operator___, and ___gene(s)___.

12. Two types of operons exist, ___induceable___ and ___repressable___.

13. Catabolic pathways are often associated with ___inducible___ (inducible/repressible) operons.

14. The erroneous incorporation of uracil instead of thymine into replicating DNA would be repaired by ___base-excision repair___

15. Negative selection is used to isolate ___auxotrophs___, microbes that have different nutritional requirements from those of wild-type microbes.

16. Genetic recombination involves the exchange of ___homologues___, two genetic elements of nearly identical nucleotide sequence.

17. Cells must be competent in order for the process of ___transformation___ to occur naturally between certain species of bacteria.

18. Conjugation requires that donor cells possess a(n) ___F-plasmid___ that codes for a(n) ___pilus___ that connects the donor to the recipient.

Short-Answer Questions for Thought and Review

1. Describe three ways in which plasmids can increase the pathogenic nature of bacteria.

2. Explain what this statement means: ". . . the information of a genotype is not always expressed as a phenotype."

3. Summarize the differences between prokaryotic and eukaryotic mRNA.

4. Explain why anticodon wobbling is useful to the cell in terms of protection against mistakes *and* the conservation of anabolic energy that would otherwise have to be spent manufacturing tRNAs.

5. Table 7.3 compares the genetic processes that occur in the cell. For each process, indicate where it occurs in prokaryotes and where it occurs in eukaryotes.

6. You are a researcher who has been given the task of determining the relative mutagenic potential of three compounds: X, Y, and Z. You set up experiments similar to those shown in Figure 7.29 on page 225. Without any mutagens, you see 5 colonies that are antibiotic resistant. Exposure of cells to X produces 6 colonies, Y produces 28, and Z produces 15. Determine the rate of mutation for each compound, and list the compounds in order of increasing mutagenic potential.

7. In the Ames test, the assumption is made that if something is mutagenic in *Salmonella*, it is likely to be mutagenic in humans also and thus should be investigated further for the ability to cause cancer in animals. What is the genetic basis for this assumption?

Critical Thinking

1. Semiconservative replication of DNA results in the production of new daughter molecules that contain one new strand paired with one original parent strand. Some RNA viruses have only a single genomic strand, not complementary strands as in dsDNA. Explain, based on this structural difference alone, why RNA viruses have a higher rate of mutation than any dsDNA genome (viral or otherwise).
2. Below is a sequence of prokaryotic DNA. Transcribe the DNA into mRNA and then into protein using Figure 7.12, which shows the genetic code.

   5' — TACAAAGAGTAGGGAGGCAGCATCGGCCAT— 3'
3. You are given the prokaryotic DNA sequence below. When you translate it into mRNA, you will see that there is no start codon, and thus the DNA

codes for "nothing." Write out the RNA sequence and then show one substitution, one insertion, and one deletion that would result in a sequence with a start codon that could be translated. Once you have done this, translate each sequence into protein.

5'— TACGCGTCTATCACG — 3'

Concept Building Questions

1. In previous chapters, we have learned about microbial structure, function, and metabolism. In this chapter we learned about microbial genetics. Ultimately, a cell is the product of its genes, but it is the phenotype, not the genotype, that is subject to evolutionary processes. Explain why it is the appearance, function, and metabolism of a microbe, rather than its actual genes, that determine whether a microbe can survive in the environment.
2. In Chapter 3, we learned about prokaryotic structure and, in particular, about transporters that move molecules across the cell membrane. Assume symporter A brings galactose into the cell along a sodium gradient. Let us assume that the symporter is encoded in an operon along with enzymes needed to catabolize galactose. Propose a mechanism that would lead to the shutdown of the operator and the removal or inactivation of the symporter upon loss of galactose from the environment.
3. In Chapter 2, we learned about chemical bonding and protein structure. Globular proteins are generally hydrophobic on the interior and hydrophilic on the surface. Indicate how a missense mutation that results in placement of a hydrophobic amino acid on the surface of the protein could change the physical structure of the protein and the way it interacts with other proteins and molecules in the surrounding environment.

8 Recombinant DNA Technology

CHAPTER SUMMARY

The Role of Recombinant DNA Technology in Biotechnology (p. 240)

Biotechnology is the use of microorganisms to make practical products such as bread, wine, paper, and antibiotics. Since the 1990s, scientists have become increasingly adept at intentionally modifying the genomes of organisms, by natural processes, for a variety of practical purposes. This **recombinant DNA technology** has expanded the possibilities of biotechnology. Scientists who manipulate genomes have three main goals:

To eliminate undesirable phenotypic traits in humans, animals, plants, and microbes.

To combine beneficial traits of two or more organisms to create valuable new organisms.

To create organisms that synthesize products that humans need.

The Tools of Recombinant DNA Technology (pp. 240–245)

Scientists use a variety of physical agents, naturally occurring enzymes, and synthetic molecules to manipulate genes and genomes and create gene libraries.

Mutagens

Mutagens are chemical and physical agents used to create changes in a microbe's genome to effect desired changes in the microbe's phenotype. For example, researchers exposed the fungus *Penicillium* to mutagenic agents and then selected strains that produce greater amounts of penicillin.

The Use of Reverse Transcriptase to Synthesize cDNA

Reverse transcriptase is an enzyme that transcribes DNA nucleotides from an RNA template. Genetic researchers use it to **make complementary DNA (cDNA)**, so called because it is complementary to an RNA template. Because eukaryotic cDNA lacks noncoding sequences, scientists can insert it into prokaryotic cells, making it possible for the prokaryotes to produce eukaryotic proteins such as human growth factor, insulin, and blood-clotting factors.

Synthetic Nucleic Acids

Computer technology has allowed for the development of a machine that synthesizes molecules of DNA and RNA with any nucleotide sequence entered onto the machine's keyboard. Scientists used synthetic nucleic acids to elucidate the genetic code, and they now use them to create genes for specific proteins. They also use them to synthesize DNA and RNA **probes**, which are nucleic acid molecules with a specific nucleotide sequence that have been labeled with radioactive or fluorescent chemicals so that their locations can be detected. Probes help locate specific DNA sequences such as genes for particular polypeptides. Finally, synthetic nucleic acids can be used to make antisense nucleic acid molecules that bind to and interfere with genes and mRNA molecules.

Restriction Enzymes

Restriction enzymes are enzymes used by bacterial cells to protect against phages by cutting phage DNA into nonfunctional pieces. Scientists use restriction enzymes to cut DNA at **restriction sites**, specific nucleotide sequences that are usually palindromic. They then combine these bits of DNA with ligase to form recombinant DNA molecules.

Vectors

In recombinant DNA technology, a **vector** is a small DNA molecule (such as a viral genome, transposon, or plasmid) that carries a particular gene and a recognizable gene marker into a cell so that the cell will develop a new phenotype—for example, the ability to synthesize growth hormone. Vectors must be small enough to manipulate in the laboratory, be able to survive inside cells, contain a recognizable genetic marker, and provide the required genetic elements.

Gene Libraries

A **gene library** is a collection of bacterial or phage clones—identical descendants—each of which carries a fragment (typically a single gene) of an organism's genome. Genetic researchers create each of the clones in a gene library by using restriction enzymes to generate fragments of the DNA of interest and then ligase to synthesize recombinant vectors. They insert the vectors into bacterial cells, which are then grown on culture media. Many gene libraries are now commercially available.

Techniques of Recombinant DNA Technology (pp. 245–249)

Mutagens, restriction enzymes, vectors, and the other tools of recombinant DNA technology are used in a variety of techniques to multiply, identify, manipulate, isolate, map, and sequence the nucleotides of genes.

Multiplying DNA *in vitro*: The Polymerase Chain Reaction

The **polymerase chain reaction (PCR)** is a technique by which scientists produce a large number of identical molecules of DNA *in vitro*. PCR is a repetitive process that alternately separates and replicates the two strands of DNA. Each cycle consists of three steps: **denaturation** (separation of the two strands of DNA), **priming** (addition of nucleotide mixture followed by cooling), and **extension** (heating to increase the rate of replication). With each repetition of the cycle, the number of DNA molecules increases exponentially. After 30 cycles in a few hours, PCR produces over a billion copies of the original DNA molecule.

Selecting a Clone of Recombinant Cells

Researchers use probes to find clones containing the DNA of interest. They then isolate and culture cells that have the desired radioactive or fluorescent marker.

Separating DNA Molecules: Gel Electrophoresis and the Southern Blot

Gel electrophoresis is a technique for separating molecules (including fragments of nucleic acids) by size, shape, and electrical charge. The technique involves drawing DNA molecules, which have an overall negative charge, through a semisolid gel by an electric current toward the positive electrode within an electrophoresis chamber. Smaller DNA fragments move faster and farther than larger ones. In genetic engineering, scientists use the technique to isolate fragments of DNA molecules that can then be inserted into vectors, multiplied by PCR, or preserved in a gene library.

The **Southern blot** technique begins with the procedures of gel electrophoresis just described, but allows researchers to stabilize specific DNA sequences and then localize them using DNA dyes or probes. The Southern blot is used for genetic fingerprinting, diagnosis of infectious disease, and other purposes.

DNA Microarrays

DNA microarrays consist of molecules of single-stranded DNA that have been immobilized on glass slides, silicon chips, or nylon membranes. Single strands of fluorescently labeled DNA in a sample washed over an array adhere only to locations on the array where there are complementary DNA sequences. DNA microarrays can be used in a number of ways, including monitoring gene expression, diagnosing infection, and identifying organisms in an environmental sample.

Inserting DNA into Cells

In addition to using vectors and the natural methods of transformation of competent cells, transduction, and conjugation, scientists have developed several artificial methods to introduce DNA into cells, including the following:

Electroporation uses an electrical current to puncture microscopic holes through a cell's membrane so that DNA can enter. Thick cell walls are first removed enzymatically to form protoplasts.

Protoplast fusion is a process in which the cytoplasmic membranes of protoplasts encountering each other fuse to form a recombinant molecule.

Injection is a process in which scientists use a gene gun to inject a cell with a tungsten or gold bead coated with DNA. In microinjection, a glass micropipette is used.

Applications of Recombinant DNA Technology (pp. 249–255)

Recombinant DNA technology has a wide range of applications.

Genetic Mapping

Genomics is the sequencing, analysis, and comparison of genomes. Genetic sequencing, called genetic mapping, involves locating genes on a nucleic acid molecule. The technique of nucleotide sequencing has been speeded up by an automated machine that distinguishes among fluorescent dyes attached to each type of nucleotide base. Scientists can also identify a particular gene using a fluorescent probe complementary to the target gene, a process called **fluorescent in situ hybridization (FISH)**. Scientists have elucidated complete gene maps of numerous viral, bacterial, and eukaryotic organisms. Another use for genomics is to relate DNA sequence data to protein function.

Environmental Studies

Unique DNA sequences (also called **signatures** or **DNA fingerprints**) reveal the presence of microbes that have never been cultured in a laboratory. These discoveries can potentially lead to a better understanding of numerous environmental and public health concerns, and help us to develop methods to resolve large-scale problems, such as global warming.

Pharmaceutical and Therapeutic Applications

Pharmaceutical and therapeutic applications of recombinant DNA technology include the following:

Protein synthesis. Scientists have inserted genes for insulin and other proteins into bacteria and yeast cells so that the microbes synthesize these proteins in vast quantities. Genetically engineered proteins are safer and less expensive than proteins isolated from donated blood or from animals.

Vaccines. Scientists synthesize subunit vaccines—which use a portion of a pathogen rather than the pathogen itself—by introducing genes for a pathogen's polypeptides into vectors. When the vectors, or the polypeptides they produce, are injected into a human, the body's immune system is exposed to and reacts against relatively harmless antigens instead of the potentially harmful pathogen.

Genetic screening. Genetic mutations cause some diseases such as inherited forms of breast cancer and Huntington's disease. In genetic screening, laboratory technicians use DNA microarrays to screen a patient's blood or other tissues for these genetic mutations before the patient shows any sign of the disease.

DNA fingerprinting. Medical laboratory technicians and forensic investigators use gel electrophoresis and Southern blotting for so-called genetic fingerprinting or DNA fingerprinting, to identify individuals or organisms by their unique DNA sequences. The technique is used in paternity investigations, crime scene forensics, diagnostic microbiology, and epidemiology.

Gene therapy. In gene therapy, missing or defective genes are replaced with normal genes, in theory curing the genetic disease. For example, patients have been successfully treated for severe combined immunodeficiency disease (SCID). Unfortunately, gene therapy has also caused patient deaths, because some patients' immune systems have hyperresponded to the presence of the vectors.

Medical diagnosis. Clinical microbiologists use PCR, fluorescent genetic probes, and DNA microarrays in diagnostic applications such as examining patient specimens for sequences unique to certain pathogens.

Xenotransplants. In xenotransplants involving recombinant DNA technology, human genes are inserted into animals to produce cells, tissues, or organs that are then introduced into the human body.

Agricultural Applications

Recombinant DNA technology has been applied in agriculture to produce **transgenic organisms**, recombinant plants and animals that have been altered for specific purposes by the addition of genes from other organisms. Agricultural uses of recombinant DNA technology include advances in herbicide tolerance, salt tolerance, freeze resistance, and pest resistance, as well as improvements in nutritional value and yield.

The Ethics and Safety of Recombinant DNA Technology (pp. 255–256)

Among the ethical and safety issues surrounding recombinant DNA technology are concerns over the accidental release of altered organisms into the environment, the ethics of altering animals for human use, and the potential for creating genetically modified biological weapons. Some opponents of recombinant DNA techniques argue that transgenic organisms could trigger allergies or cause harmless organisms to become pathogenic, whereas others caution that the long-term effects of transgenic manipulations are unknown and that unforeseen problems arise from every new technology. In addition, emergent recombinant DNA technologies raise numerous other ethical issues, debating the rights of governments, employers, or insurers to routinely screen people for certain diseases, the rights of individuals to refuse either screening or treatment for genetic disease, and the rights of both institutions and individuals to privacy and confidentiality of genetic data. Our society will have to confront these and other ethical issues as the genomic revolution continues.

KEY THEMES

Microbes are extremely versatile organisms. They thrive everywhere and consume just about everything. They are also, in many cases, easy to propagate in the laboratory. The field of recombinant DNA technology makes use of both the natural tendencies of microbes and our ability to manipulate them to enhance our human experience. Remember the following as you study this chapter:

> *The creation of recombinant DNA involves more than just molecular biology:* Molecular biology allows us to change DNA almost at will, but we always need to remember the environmental, evolutionary, and ethical implications of such changes.

QUESTIONS FOR FURTHER REVIEW

Answers to these questions can be found in the answer section at the back of this study guide. Refer to the answers only after you have attempted to solve the questions on your own.

Multiple Choice

1. The enzyme reverse transcriptase
   a. was discovered in retroviruses.
   b. uses an RNA template to make DNA.

c. is used to make cDNA.
d. All of the above are true about reverse transcriptase.

2. In nature, the normal function of restriction enzymes is
   a. to facilitate horizontal gene transfer.
   b. to remove mismatched nucleotides during DNA repair.
   c. to protect the cell from phage DNA by cutting it up.
   d. to remove old cellular DNA by cutting it up.
3. Which of the following elements must be present in a recombinant vector to allow selection for the presence of the vector and its new gene in a host cell?
   a. specific restriction sites
   b. genetic marker
   c. promoter
   d. origin of replication
4. Which molecule listed below is NOT used in PCR?
   a. DNA template
   b. DNA polymerase
   c. diphosphate deoxynucleotides
   d. DNA primers
5. Which of the methods below of introducing DNA into cells can occur naturally as well as artificially?
   a. protoplast fusion
   b. transformation
   c. conjugation
   d. electroporation
6. Which of the following is a true statement about genomics?
   a. Genomics involves the sequencing and analysis of the human genome.
   b. Genomics involves the sequencing and analysis of a variety of genomes.
   c. Genomics employs time-consuming techniques to sequence DNA.
   d. Genomics provides sequences of DNA, but little other information.
7. Which technique below would have the most application in diagnosing human diseases?
   a. transformation
   b. xenotransplantation
   c. genetic screening
   d. DNA fingerprinting

Fill in the Blanks

1. Complementary DNA (cDNA) is useful to scientists working with eukaryotic genes because all of the introns have been removed from the sequence. Prokaryotes are not able to remove these elements, and so otherwise they would not be able to express the DNA.
2. Restriction enzymes cleave dsDNA to produce fragments with either Sticky ends or blunt ends. DNA molecules are more easily combined with Sticky ends because of complementarity.
3. The technique used to amplify small amounts of DNA in vitro is called PCR.

4. Gel electrophoresis is a technique that allows the separation of molecules based on three parameters: size, shape, and charge.

5. Southern blots transfer DNA from agarose gels to membranes, whereas northern blots transfer RNA.

6. DNA fingerprinting uses gel electrophoresis and Southern blotting to identify individuals based on their unique DNA sequences.

7. A(n) transgenic organism is an organism that has been specifically altered by the insertion of genetic elements from a different organism.

Short-Answer Questions for Thought and Review

1. Summarize the three major goals of recombinant DNA technology. In one sentence, state why these goals are ethically controversial.

2. Contrast the overall application of recombinant DNA technology to human medicine and to agriculture. What is the central medical focus of the technology? What is the central agricultural focus?

3. List at least three benefits of recombinant DNA technology and at least three potential hazards. Your answers do not have to relate solely to humans.

Critical Thinking

1. You are given the tripeptide sequence below. Back-translate the protein sequence and write all possible mRNAs and cDNAs that could give rise to this single peptide. Figure 7.12 may be useful.

$$NH_2 — Met — Ala — Phe — COOH$$

Concept Building Questions

1. In the Critical Thinking question above, you were asked to derive all of the mRNA and cDNA sequences that could give rise to a single tripeptide. In past chapters, we have learned about codon wobbling, classification of microbes

based on genetic sequences, and nucleic acid structure and bonding. Based on the information you have gained so far, answer the questions below.

a. How does codon wobbling make classification of microbes by sequence analysis potentially problematic?
b. Is there enough similarity in the cDNAs you created in the Critical Thinking question to positively relate the DNAs as being from the same organism? Does the chance of making a positive correlation increase with the length of the sequence? Why or why not?
c. Based on bonding, which of the cDNAs produced in the Critical Thinking question would have very stable dsDNA (high degree of bonding, high melting temperature)? Which would have low stability?

9 Controlling Microbial Growth in the Environment

CHAPTER SUMMARY

Basic Principles of Microbial Control (pp. 262–264)

In discussions about microbial control in the environment, precise terminology is important, as is an understanding of the concept of microbial death rate and the action of antimicrobial agents.

Terminology of Microbial Control

Many terms of microbial control, though familiar to the general public, are often misused. Precise definitions are as follows:

In its strictest sense, **sterilization** refers to the removal or destruction of all microbes, including viruses and bacterial endospores, in or on an object. (The term does not apply to prions.) In practical terms, sterilization techniques eradicate harmful microbes, but some innocuous microbes may still be present.

The term **aseptic** describes an environment or procedure that is free of contamination by pathogens.

Disinfection refers to the use of physical or chemical agents known as *disinfectants* to inhibit or destroy microorganisms, especially pathogens. It does not guarantee elimination of all pathogens and applies only to treatment of inanimate objects.

When a chemical is used on skin or other tissue, the process is called **antisepsis** and the agent is an **antiseptic**.

Degerming is the removal of microbes from a surface by scrubbing, whether that surface is human skin or a table top.

Sanitization is the process of disinfecting plates and utensils used by the public to reduce the number of pathogenic microbes so as to meet acceptable public health standards. Dishes are disinfected at home but are sanitized in a restaurant.

Pasteurization is the use of heat to kill pathogens and reduce the number of spoilage microorganisms in food and beverages. Milk, fruit juices, wine, and beer are commonly pasteurized.

Agents or techniques that inhibit the growth of microbes without necessarily killing them are indicated by the suffix *-stasis* or *-static*. For example, refrigeration is *bacteriostatic*. By contrast, words ending in *-cide* or *-cidal* refer to agents or methods that destroy or permanently inactivate a particular type of microbe. For example, *fungicides* kill fungal hyphae, spores, and yeasts.

Microbial Death Rates

Scientists define **microbial death** as the permanent loss of reproductive ability under ideal environmental conditions. One technique for evaluating the efficacy of an antimicrobial agent is to calculate the **microbial death rate**, which is usually found to be constant over time for any particular microorganism under a particular set of conditions. When the microbial death rate is plotted on a semilogarithmic graph, this constant death rate produces a straight line.

Action of Antimicrobial Agents

The modes of action of antimicrobial agents fall into two basic categories: those that disrupt the integrity of cells by adversely altering their cell walls or cytoplasmic membranes, and those that interrupt cellular metabolism and reproduction by interfering with the structures of proteins and nucleic acids.

The Selection of Microbial Control Methods (pp. 264–267)

A perfect antimicrobial method or agent would be inexpensive, fast-acting, stable during storage, harmless to humans, and effective against all types of microbes. Because no single method or agent meets all these critera, scientists consider several factors when evaluating methods and agents.

Factors Affecting the Efficacy of Antimicrobial Methods

One factor affecting the choice of antimicrobial is the site to be treated. For example, harsh chemicals or intense heat cannot be used on human tissues. Another factor is the relative susceptibility of the microorganisms. Generally, scientists and medical personnel select a method to kill the hardiest microorganisms present, assuming that more fragile microbes will be killed as well. The most resistant microbes are bacterial endospores, species of *Mycobacterium*, and cysts of protozoa. However, infectious proteins, called prions, are more resistant than any living thing. The third factor in antimicrobial efficacy is the environmental conditions under which it is used, such as temperature and pH. For example, because chemicals react faster at higher temperatures, warm disinfectants generally work better than cool ones.

Methods for Evaluating Disinfectants and Antiseptics

Scientists have developed several methods to measure the efficacy of antimicrobial agents:

> Phenol was an antiseptic used during surgery in the late 1800s. Since then, scientists have evaluated the efficacy of various disinfectants and antiseptics by calculating a ratio that compares the agent's ability to control microbes to that of phenol. This ratio is referred to as the **phenol coefficient**. A phenol coefficient greater than 1.0 indicates that an agent is more effective than phenol. In the **use-dilution test**, a researcher dips several metal cylinders into broth cultures of bacteria, briefly dries them, then immerses each into a different dilution of the disinfectants being evaluated. After 10 minutes, the cylinders are removed and incubated. The most effective agent is the one that entirely prevents microbial growth at the highest dilution.

The **Kelsey-Sykes capacity test** is an alternative to the use-dilution test that is the standard in the European Union to determine the capacity of a given chemical to inhibit bacterial growth.

In-use tests provide accurate determination of an agent's efficacy under environmental conditions, such as when swabs are taken from objects in a hospital emergency department.

Physical Methods of Microbial Control (pp. 267–275)

Physical methods of microbial control include exposing the microbes to extremes of heat and cold, desiccation, filtration, osmotic pressure, and radiation.

Heat-Related Methods

Heat is one of the older and more common means of microbial control. High temperatures denature proteins, interfere with the integrity of cytoplasmic membranes and cell walls, and disrupt the function and structure of nucleic acids. Microorganisms vary in their susceptibility to heat. The **thermal death point** is the lowest temperature that kills all cells in a broth in 10 minutes, and **thermal death time** is the time it takes to completely sterilize a particular volume of liquid at a set temperature. **Decimal reduction time** (D) is the time required to destroy 90% of the microbes in a sample.

Moist heat is more effective than dry heat because water is a better conductor of heat than air. Boiling kills the vegetative cells of bacteria and fungi, the trophozoites of protozoa, and most viruses within 10 minutes at sea level. It is not effective when true sterilization is required. In such cases, autoclaving is required. An **autoclave** is a device consisting of a pressure chamber, pipes, valves, and gauges that uses steam heat under pressure to sterilize chemicals and objects that can tolerate moist heat.

Pasteurization, a method of heating foods to kill pathogens and control spoilage organisms without altering the quality of the food, can be achieved by several methods: the historical (batch) method at 63°C for 30 minutes, flash pasteurization at 72°C for 15 seconds, and ultrahigh-temperature pasteurization at 134°C for 1 second.

Substances that cannot be sterilized by moist heat, such as powders and oils, can be sterilized using dry heat at much higher temperatures for longer times. Complete incineration is the ultimate means of sterilization.

Refrigeration and Freezing

Refrigeration (temperatures between 0°C and 7°C) halts the growth of most pathogens, which are predominantly mesophiles. Slow freezing at temperatures below 0°C is effective in inhibiting microbial metabolism; however, many vegetative bacterial cells, bacterial endospores, and viruses can survive subfreezing temperatures for years.

Desiccation and Lyophilization

Desiccation, or drying, has been used for thousands of years to preserve such foods as fruits, peas, and yeast. It inhibits microbial growth because metabolism requires liquid water. **Lyophilization**, or freeze-drying, preserves microbes and other cells for many years. In this process, scientists freeze a culture in liquid nitrogen or frozen carbon dioxide and then remove the water via a vacuum. Lyophilization prevents the formation of large ice crystals, which would otherwise damage the culture; thus, enough viable cells remain to enable the culture to be reconstituted many years later.

Filtration

When used as a method of microbial control, **filtration** is the passage of air or a liquid through a material that traps and removes microbes. Some membrane filters manufactured of nitrocellulose or plastic have pores small enough to trap the smallest viruses and even some large protein molecules. *HEPA (high-efficiency particulate air)* filters remove microbes and particles from air.

Osmotic Pressure

High concentrations of salt or sugar inhibit microbial growth by **osmotic pressure**, drawing out of cells the water they need to carry out their metabolic functions. Honey, jams, salted fish, and pickles are examples of foods preserved by osmotic pressure. Fungi have a greater tolerance for hypertonic environments than do bacteria, which explains why refrigerated jams may grow mold.

Radiation

There are two types of **radiation:** *Particulate radiation* consists of high-speed subatomic particles freed from their atoms, whereas *electromagnetic radiation* is atomic energy without mass traveling at the speed of light. **Ionizing radiation** is electromagnetic radiation with wavelengths shorter than 1 nm, such as electron beams, gamma rays, and X rays. It creates ions that produce effects leading to the denaturation of important molecules and cell death. **Nonionizing radiation**, such as ultraviolet light, visible light, infrared light, and radio waves, has wavelengths longer than 1 nm. Of these types, only ultraviolet light has sufficient energy to be a practical antimicrobial agent. It causes pyrimidine dimers, which can kill affected cells.

Biosafety Levels

The Centers for Disease Control and Prevention has established guidelines for four levels of safety in microbiological laboratories dealing with pathogens. Biosafety Level 1(BSL-1) is suitable for handling microbes not known to cause disease in healthy humans. BSL-2 facilities are equipped to handle moderately hazardous agents and require precautions to limit exposure to aerosols or contaminated objects. BSL-3 labs require all manipulations be done in safety cabinets and have special design features to control the flow of air through the room. BSL-4 facilities are for work with microbes that cause severe or fatal disease in humans. These facilities are isolated, and access is strictly controlled. Personnel must wear suits with air hoses to protect them from exposure to the microbes.

Chemical Methods of Microbial Control (pp. 275–281)

Nine major categories of antimicrobial chemicals are used as antiseptics and disinfectants.

1. Phenol and Phenolics

Phenolics are compounds derived from phenol molecules that have been chemically modified by the addition of halogens or organic functional groups such as chlorine. They are intermediate- to low-level disinfectants that denature proteins and disrupt cell membranes in a wide variety of pathogens.

2. Alcohols

Alcohols such as isopropanol (rubbing alcohol) denature proteins and disrupt cell membranes; they are used either as 70–90% aqueous solutions or in a *tincture*, which is a combination of an alcohol and another antimicrobial agent. Alcohols are bactericidal, fungicidal, and virucidal against enveloped viruses; however, they are not effective against fungal spores or bacterial endospores. They are considered intermediate-level disinfectants.

3. Halogens

Halogens are the four very reactive, nonmetallic chemical elements: iodine, chlorine, bromine, and fluorine. Halogens are used as intermediate-level disinfectants and antiseptics to kill microbes in water or on medical instruments or skin. Although their exact mode of action is unknown, they are believed to denature enzymes. Iodine is used medically, whereas chlorine is more commonly used by municipalities to treat drinking water supplies, wastewater, and swimming pools. In hot tubs, bromine is more effective than chlorine, because it evaporates more slowly at high temperatures. Fluorine in the form of fluoride is an antibacterial added to drinking water and toothpaste to help reduce the incidence of dental caries.

4. Oxidizing Agents

Oxidizing agents such as hydrogen peroxide, ozone, and peracetic acid are high-level disinfectants and antiseptics that release oxygen radicals, which are toxic to many microbes, especially anaerobes. Hydrogen peroxide can disinfect and even sterilize surfaces, but it is not useful in treating open wounds, because it is quickly neutralized by the catalase enzyme released from damaged human cells. Some Canadian and European municipalities use ozone rather than chlorine to treat their drinking water because ozone is more effective as an antimicrobial and does not produce carcinogenic by-products. Peracetic acid is an extremely effective sporicide used to sterilize equipment.

5. Surfactants

Surfactants are "surface active" chemicals. They include *soaps*, whose molecules have both hydrophobic ends, which act primarily to break up oils during degerming, and negatively charged hydrophilic ends, which attract water. *Detergents* are positively charged organic surfactants that disrupt cellular membranes. **Quaternary ammonium compounds (quats)** are an example. However, quats are considered low-level disinfectants because they are not effective against mycobacteria, endospores, or nonenveloped viruses, and some pathogens actually thrive in them.

6. Heavy Metals

Heavy metal ions such as arsenic, silver, mercury, copper, and zinc are low-level disinfectants that denature proteins. For most applications, they have been superceded by less toxic alternatives, but silver still plays an antimicrobial role in some surgical dressings, burn creams, and catheters.

7. Aldehydes

Aldehydes are compounds containing terminal —CHO groups. Classified as high-level disinfectants, they cross-link organic functional groups in proteins and DNA. A 2% solution of glutaraldehyde and a 37% aqueous solution of formaldehyde (called *formalin*) are used to disinfect or sterilize medical or dental equipment and in embalming fluid.

8. Gaseous Agents

Many items, such as plastic laboratory ware, artificial heart valves, mattresses, and dried foods, cannot be sterilized easily with heat or water-soluble chemicals, nor is irradiation always practical. However, such items can be sterilized within a closed chamber containing highly reactive gases such as *ethylene oxide*, *propylene oxide*, and *beta-propiolactone*, which denature proteins and DNA by cross-linking organic functional groups, thereby killing everything they contact without harming inanimate objects. These gases are explosive and potentially carcinogenic.

9. Enzymes

Antimicrobial enzymes are enzymes that act against microorganisms. Scientists, food processors, and medical personnel are researching ways to use natural and chemically modified antimicrobial enzymes to control microbes in the environment, inhibit microbial decay of foods and beverages, and reduce the number and kinds of microbes on medical equipment. One exciting development is the use of an enzyme to eliminate the prion that causes variant Creutzfeldt-Jakob disease.

Antimicrobials

Antimicrobials include *antibiotics*, which are produced naturally by microorganisms; *semisynthetics*, which are chemically modified antibiotics; and *synthetics*, which are wholly synthetic antimicrobial drugs. These compounds are typically used to treat disease but can also function as intermediate-level disinfectants.

Development of Resistant Microbes

There is little evidence that the extensive use of antimicrobial chemicals in household cleansers and personal care products enhances human health; however, it does promote the development of strains of microbes resistant to antimicrobial chemicals. The reason is that when susceptible cells die, they reduce competition for resources, allowing any remaining resistant cells to proliferate. Many experts therefore recommend limiting the use of such chemicals to food handling and health care involving high-risk patients and newborns.

KEY THEMES

As we have seen so far in our studies, microbes are very durable organisms, capable of surviving almost everywhere. In nature, this is fine, but this hardiness becomes extremely problematic when we try to confront microbes that are capable of making us sick. As you study this chapter, focus on this key observation:

> *Microbial control is one of the fundamental goals of microbiology:* By intensive study, we have learned much about microorganisms, their structure, and how they survive. This knowledge helps us to better understand

the biological world and provides us with the tools we need to develop physical and chemical methods of controlling microbial growth around us, on us, and inside us.

QUESTIONS FOR FURTHER REVIEW

Answers to these questions can be found in the answer section at the back of this study guide. Refer to the answers only after you have attempted to solve the questions on your own.

Multiple Choice

1. Which of the following would NOT be destroyed by sterilization?
   a. bacteria
   b. prions
   c. viruses
   d. All could be destroyed by sterilization.
2. Which microbial control method below could be used directly on a hospital patient?
   a. degerming
   b. sanitization
   c. antisepsis
   d. both a and c
3. Assume that a chemical control agent causes missense mutations within the bacterial population it is used against. The agent is
   a. bactericidal.
   b. bacteriostatic.
   c. either bactericidal or bacteriostatic.
   d. neither bactericidal nor bacteriostatic.
4. Which of the following evaluation techniques is no longer used to gauge the effectiveness of antiseptics?
   a. phenol coefficient
   b. use-dilution test
   c. in-use test
   d. All the above are still being used.
5. Which test below is actually used to determine the effectiveness of chemicals in the environment?
   a. phenol coefficient
   b. use-dilution test
   c. in-use test
   d. none of the above
6. Endospores are least likely to survive which heating method?
   a. boiling
   b. autoclaving
   c. pasteurization
   d. Endospores can routinely survive all of these methods.
7. Moist and dry heat generally work to kill microbes by
   a. physically destroying/denaturing cellular proteins.
   b. physically incinerating the cells.
   c. physically denaturing nucleic acids.
   d. inhibiting cellular processes by allosterically modifying proteins.
8. Which method below would allow for the most effective preservation of viable microbial samples over long periods of storage?
   a. slow freezing
   b. desiccation
   c. lyophilization
   d. refrigeration

9. Assume you need to remove all of the microbes from a very large volume of liquid. Which method below would be the most effective and efficient in doing this?
   a. autoclaving
   b. filtration
   c. radiation
   d. pasteurization
10. Which of the following physical methods of microbial control is NOT used in the food industry?
    a. pressure cooking
    b. pasteurization
    c. osmotic pressure
    d. nonionizing radiation
11. Chemical methods of microbial control would be LEAST effective against which microbial entity?
    a. enveloped virus
    b. fungal cell
    c. bacterial endospore
    d. nonenveloped virus
12. Which chemical control agent listed below works against the broadest spectrum of microbes?
    a. phenols
    b. alcohols
    c. halogens
    d. oxidizing agents
13. Soaps, by themselves, are examples of
    a. surfactants.
    b. degerming agents.
    c. antiseptics.
    d. both a and b.
14. Which chemical control agent listed below has the highest disinfectant classification?
    a. heavy metals
    b. halogens
    c. aldehydes
    d. phenolics
15. Which chemical control agent listed below is NOT an antiseptic?
    a. ethylene oxide
    b. detergent
    c. rubbing alcohol
    d. iodine

Fill in the Blanks

1. Bleach is an example of a chemical ____________________.
2. If an antibiotic kills a bacterial pathogen, it is said to be ____________________; if it only inhibits the growth of the bacterial pathogen, it is said to be ____________________.
3. Temperature generally ____________________ (increases/decreases) the effectiveness of antimicrobial agents.
4. ____________________ is the lowest temperature that will kill all of the cells (sterilize) in a broth culture in 10 minutes.
5. ____________________ heat is more effective in killing microbes than ____________________ heat.

6. ____________________ and ____________________ bacteria are capable of surviving pasteurization, but they are generally not pathogenic and so don't make us sick even if they remain in food or beverages.
7. The two types of radiation that can be used to control microbial populations are ____________________ and ____________________.
8. Triclosan is a(n) ____________________ found in many consumer products such as hand soaps.
9. Iodine is used in two ways medically, either as a(n) ____________________ or as a(n) ____________________.
10. ____________________ are antimicrobial chemicals that are made naturally by certain microorganisms.

Short-Answer Questions for Thought and Review

1. List the following microbial control techniques in order of effectiveness from least effective at removing microbes to most effective: pasteurization, degerming, sterilization, disinfection, sanitization.
2. Summarize the two modes of action that chemical and physical microbial control agents/processes use against cells. What, in general, happens in each, and why is this a problem for the microbes?
3. What properties do chemical control agents with high activities have that make them effective?

Critical Thinking

1. A nursing home wishes to test various disinfectants to find the ones that are most effective. Which evaluation method do you think would be most efficacious in identifying the disinfectant that works best against the microorganisms in the nursing home, and why?
2. Which methods of microbial control inhibit microbial growth by effectively removing water from the microbes? Describe specifically how each method works.

3. Though there is no firm proof yet that antimicrobial soaps are leading to increases in resistance among microbes, the CDC still recommends limiting the use of these agents. Why?

Concept Building Questions

1. Figure 9.2 presents the relative susceptibilities of microbes to antimicrobial drugs. Explain, using what you learned about structure in Chapter 3, the fundamental structural differences between the more susceptible microbes and the least susceptible microbes. What specifically must a high-level germicide do to kill the most resistant microbes?
2. Based on what you have learned in previous chapters about chemical bonding, microbial structure, function, metabolism, and genetics, list some possible mechanisms by which a bacterial cell could become resistant to a chemical such as triclosan. Give an example from each topic listed in this question.

10 Controlling Microbial Growth in the Body: Antimicrobial Drugs

CHAPTER SUMMARY

The History of Antimicrobial Agents (pp. 288–289)

Chemicals that affect physiology in any manner are called *drugs*. **Chemotherapeutic agents** are drugs used to treat diseases. Among them are **antimicrobial agents (antimicrobials)**, which are drugs used to treat infections. These include **antibiotics,** which are biologically produced agents, **semisynthetics** (chemically modified antibiotics), and **synthetics** (antimicrobials that are completely synthesized in a laboratory).

The development of these drugs began with German scientist Paul Ehrlich (1854–1915), who produced arsenic compounds effective against trypanosome parasites and syphilis. In 1929, British bacteriologist Alexander Fleming (1881–1955) developed the earliest form of penicillin, and in 1932, German chemist Gerhard Domagk (1895–1964) developed the first practical antimicrobial agent, sulfanilamide.

Mechanisms of Antimicrobial Action (pp. 289–297)

The key to successful chemotherapy against microbes is **selective toxicity**; that is, an effective antimicrobial agent must be more toxic to a pathogen than to the pathogen's host. Selective toxicity is possible because of differences in structure or metabolism between the pathogen and its host. Because there are many differences between the structure and metabolism of pathogenic bacteria and their eukaryotic hosts, antibacterial drugs are the most numerous and diverse antimicrobial agents. Fewer drugs are available against pathogenic fungi, protozoa, and helminths because these are eukaryotes. Antiviral drugs are also limited, because viruses utilize their host cells' enzymes and ribosomes to metabolize and replicate. Therefore, drugs that are effective against viral replication are likely toxic to the host as well.

Inhibition of Cell Wall Synthesis

The most common antibacterial agents act by preventing the cross-linkage of NAM subunits in the bacterial cell wall, which protects a cell from the effects of osmotic pressure. Most prominent among these drugs are the **beta-lactams,** such as penicillins, cephalosporins, and monobactams. Their functional lactam rings irreversibly bind to the enzymes that cross-link NAM subunits.

Vancomycin and **cycloserine** disrupt cell wall formation by interfering with alanine-alanine bridges that link the NAM subunits in many Gram-positive bacteria. **Bacitracin** also disrupts cell wall formation by blocking the secretion of NAG and NAM from the cytoplasm to the cell wall.

Isoniazid (INH) and **ethambutol** block mycolic acid synthesis in the walls of mycobacteria, and they are used to treat tuberculosis and leprosy.

Inhibition of Protein Synthesis

Many antimicrobials take advantage of the differences between prokaryotic and eukaryotic ribosomes to target bacterial protein translation. Antimicrobials that inhibit protein synthesis include **aminoglycosides** such as streptomycin and gentamicin, and the **tetracyclines**, all of which inhibit functions of the 30S ribosomal subunit. In contrast, chloramphenicol, lincosamides, streptogramins, and the **macrolides** such as erythromycin inhibit 50S subunits. Antisense nucleic acids are designed to be complementary to specific RNA molecules of pathogens and thus block protein synthesis.

Disruption of Cytoplasmic Membranes

Some antimicrobial agents, such as the **polyenes**, disrupt the cytoplasmic membrane of a targeted cell, often by forming a channel through the membrane and damaging its integrity. The polyenes *nystatin* and *amphotericin B* are fungicidal because they attach to *ergosterol*, a lipid in fungal membranes. Azoles and allyamines are antifungal drugs that inhibit the synthesis of ergosterol. *Polymyxin* is effective against Gram-negative bacteria, but it is toxic to human kidneys and therefore is reserved for use against external pathogens that are resistant to other antibacterial drugs. Some antiparasitic drugs also act against the cytoplasmic membrane.

Inhibition of Metabolic Pathways

Antimetabolic agents target differences in the metabolic processes of a pathogen and its host. **Sulfonamides** are **structural analogs** of para-aminobenzoic acid (PABA), a compound that is crucial in DNA and RNA synthesis in some microorganisms, but not in humans. The substitution of sulfonamides in the metabolic pathway leading to nucleic acid synthesis kills those organisms. Trimethoprim is another antimetabolite that blocks this pathway. *Amantadine*, *rimantadine*, and weak organic bases are used as antiviral drugs because they can neutralize the acidic environment of phagolysosomes and thereby prevent viral uncoating. Protease inhibitors, when used as part of a "cocktail" of drugs, have revolutionized the treatment of AIDS patients in industrialized countries by interfering with the action of protease, an enzyme needed by HIV during its replication cycle.

Inhibition of Nucleic Acid Synthesis

Several drugs function by blocking either the replication of DNA or its transcription into RNA. **Nucleotide or nucleoside analogs** are incorporated into the DNA or RNA of pathogens, where they distort the shapes of the nucleic acid molecules and prevent further replication, transcription, or translation. They are often used against viruses, which are far more likely than host cells to incorporate them, and which divide more rapidly. They are also used against rapidly dividing cancer cells.

The synthetic *quinolones* and *fluoroquinolones* are active against prokaryotic DNA specifically because they inhibit DNA *gyrase*, an enzyme necessary for correct coiling of replicating bacterial DNA. Reverse transcriptase inhibitors act against reverse transcriptase, an enzyme that HIV uses in its replication cycle. Humans lack reverse transcriptase, so these inhibitors do not harm patients.

Prevention of Virus Attachment

Attachment of viruses to host cells can be blocked by peptide and sugar analogs of either attachment or receptor proteins. The use of such attachment antagonists is a new area of antimicrobial drug development.

Clinical Considerations in Prescribing Antimicrobial Drugs (pp. 297–301)

The ideal antimicrobial agent would be readily available, inexpensive, chemically stable, easily administered, nontoxic and nonallergenic, and selectively toxic against a wide range of pathogens.

Spectrum of Action

The number of different kinds of pathogens a drug acts against is known as its **spectrum of action**. Drugs that work against only a few kinds of pathogens are **narrow-spectrum drugs**, whereas those that are effective against many different kinds of pathogens are **broad-spectrum drugs**.

Efficacy

To ascertain the efficacy of antimicrobials, microbiologists conduct a variety of tests, including the following:

- A **diffusion susceptibility test (Kirby-Bauer test)** is a simple, inexpensive test widely used to reveal which drug is most effective against a particular pathogen. The procedure involves uniformly inoculating a Petri plate with a standardized amount of the pathogen in question and arranging on the plate disks soaked in the drugs to be tested. In general, the larger the **zone of inhibition** around a disk, the more effective the drug.
- A test of the **minimum inhibitory concentration** (**MIC**) attempts to quantify the smallest amount of a drug that will inhibit a pathogen. This is typically determined by either a **broth dilution test**, in which a standardized amount of a bacterium is added to serial dilutions of antimicrobial agents in tubes or wells containing broth, or an **Etest**, in which a plastic strip containing a gradient of the antimicrobial agent being tested is placed on a plate inoculated with the pathogen of interest.
- A **minimum bactericidal concentration** (**MBC**) **test** is similar to the MIC test; samples taken from clear MIC tubes are transferred to plates containing a drug-free growth medium and monitored for bacterial replication.

Routes of Administration

For external infections, topical or local administration is effective. For internal infections, drugs are administered orally, intramuscularly, or intravenously.

Safety and Side Effects

Physicians must consider the possibility of adverse side effects of chemotherapy. Toxicity is especially important to consider for pregnant women, because many drugs that are safe for adults can have adverse effects on a fetus. Some drugs trigger allergic immune responses in sensitive patients. Although relatively rare, such reactions can be life threatening, especially when they result in anaphylactic shock. Many drugs, especially broad-spectrum antibiotics, also disrupt the normal antagonism between benign normal microbiota and opportunistic pathogens and thus prompt secondary infections, such as vaginitis or thrush.

Resistance to Antimicrobial Drugs (pp. 301–313)

The Development of Resistance in Populations

Bacterial cells acquire resistance to antibiotics in two ways: through new mutations of chromosomal genes, or by acquiring resistance genes on extrachromosomal pieces of DNA called **R-plasmids** via the processes of transformation, transduction, or conjugation. When an antimicrobial agent is present and the majority of cells die, the resistant cells continue to grow and multiply, soon replacing the sensitive cells as the majority of the population.

Mechanisms of Resistance

Microorganisms may resist a drug by one or more of the following seven mechanisms:

1. Producing enzymes such as **beta-lactamase** that deactivate the drug
2. Inducing changes in the cell membrane that prevent entry of the drug
3. Altering the drug's receptor to prevent its binding
4. Altering the cell's metabolic pathways
5. Pumping the antimicrobial out of the cell
6. Growing within a biofilm to slow cellular metabolism
7. Protecting the target of an antimicrobial drug to inhibit its binding

Multiple Resistance and Cross Resistance

Pathogens can acquire resistance to more than one drug at a time, especially when resistance is conferred by R-plasmids. Pathogens that are resistant to most antimicrobial agents are sometimes called *superbugs*, and they pose unique problems to health care professionals. **Cross resistance** typically occurs when drugs are similar in structure.

Retarding Resistance

Resistance can be retarded in one or more of four ways:

1. Using sufficiently high concentrations of a drug for a sufficient time to kill all sensitive cells and inhibit others long enough for the body's defenses to destroy them
2. Using antimicrobials in combination, promoting **synergism**, the interplay between drugs that results in efficacy that exceeds the efficacy of either drug alone
3. Limiting the use of antimicrobials to necessary cases, avoiding indiscriminate prescribing and uncontrolled use
4. Developing new variations of existing drugs

Particular use of antimicrobial drugs against specific pathogens is covered in relevant chapters.

KEY THEMES

In Chapter 9 we described methods to control the growth of microbes in the environment. Here, we consider chemotherapeutic agents—natural, semisynthetic, and synthetic chemicals that can be used in the human body to control the growth of pathogens. Fundamental to treating pathogenic diseases is the following:

> *The best drug target is one that is found in the microbe but not in the human:* Antimicrobial agents are, by default, poisons designed to kill or inhibit the growth of microbes. If the drugs are not specific enough, they will kill the host as well. Finding specific targets, and then making a drug to match, is one of the biggest challenges microbiologists and chemists face in the war against infectious diseases.

QUESTIONS FOR FURTHER REVIEW

Answers to these questions can be found in the answer section at the back of this study guide. Refer to the answers only after you have attempted to solve the questions on your own.

Multiple Choice

1. Antibiotics kill or inhibit the growth of
   a. bacteria.
   b. fungi.
   c. viruses.
   d. all of the above.
2. Chemotherapeutic agents should be
   a. more toxic to the pathogen than to the host.
   b. readily absorbed to reach appropriate levels in the host.
   c. stable so they don't form toxic compounds upon degradation.
   d. all of the above.
3. Which of the following antibacterial agents will work against *Mycobacterium*?
   a. penicillin
   b. bacitracin
   c. ethambutol
   d. vancomycin
4. Ribosomal inhibitors can work against
   a. bacterial ribosomes.
   b. mitochondrial ribosomes.
   c. fungal ribosomes.
   d. both a and b.
5. Amantadine is used to prevent infection of influenza A; it does so by targeting
   a. viral attachment to the host.
   b. viral uncoating.
   c. viral replication.
   d. viral exit from the host cell.
6. Attachment antagonists, though still experimental, would be most useful against
   a. bacteria.
   b. fungi.
   c. protozoa.
   d. viruses.
7. Which of the drugs below has the broadest spectrum of action?
   a. amphotericin B
   b. AZT
   c. amoxicillin
   d. streptomycin
8. Which list below gives the best indication of the order in which infections in a given organ system can be successfully treated (first = best success, last = least success)?
   a. brain, heart, lungs, skin
   b. heart, brain, lungs, skin
   c. skin, lungs, heart, brain
   d. lungs, skin, heart, brain

9. Which mechanism below does NOT lead to resistance in microorganisms?
   a. The microbe can pump the drug out of the cell.
   b. The microbe can alter porins to prevent entry of the drug.
   c. The microbe can remove all porins from the cell surface to prevent entry of the drug.
   d. The microbe can produce an enzyme to destroy the drug.
10. Which of the following statements is true about synergism?
   a. Synergism describes the relationship between pathogens and normal microbiota.
   b. Synergism describes how one drug can enhance the effect of another drug.
   c. Synergism can be used to prevent or slow the process of acquired resistance among microbes.
   d. Both b and c are true.
11. Which microbes produce the most naturally occurring antibiotics?
   a. bacteria
   b. fungi
   c. protozoa
   d. viruses
12. According to Table 10.2, the most common target of antibacterial drugs is
   a. the cell wall.
   b. metabolism.
   c. cytoplasmic membranes.
   d. protein synthesis.

Fill in the Blanks

1. The majority of antibiotics and semisynthetics are obtained from species of ____________________.
2. Chemotherapeutic agents that target bacterial cell wall synthesis work only on cells that are ____________________.
3. ____________________, a polyene, is effective against ____________________ (name the organism); it binds to and disrupts the membrane to cause lysis.
4. The first antimicrobial agent that was commercially available was ____________________.
5. A drug that kills or inhibits a limited number of pathogens is said to be a(n) ____________________ drug, whereas drugs that are effective against many pathogens are ____________________.
6. Microbes ____________________ (can/cannot) acquire resistance to more than one drug at a time.

7. Resistant microbes are frequently found in __________________ and __________________ because of constant antimicrobial use in these places.
8. __________________ describes what happens when resistance to one drug leads to resistance to similar drugs.

Short-Answer Questions for Thought and Review

1. Discuss why there are so few antiviral agents relative to antibacterial agents.
2. Why are drugs such as actinomycin that target DNA synthesis generally not prescribed to treat human infections?
3. List the factors that clinicians must consider before prescribing a chemotherapeutic agent for a patient. Which factor do you think carries the most weight in their decisions?

Critical Thinking

1. Of the six categories of inhibition listed in the text and shown in Figure 10.2, which would be effective against viruses? Of these, which would actually be practical to use?
2. You go to the doctor because you have a bacterial infection in your upper respiratory tract. The doctor prescribes a semisynthetic erythromycin derivative for you to take orally for 10 days. The bacteria in your lungs die, but so do the *E. coli* in your intestinal tract. Why?
3. Why is an MBC test useful to perform when trying to determine which drug to prescribe to a patient and for how long?
4. Explain how normal microbiota can protect a host by antagonism.
5. Some strains of HIV are resistant even to "cocktails" of multiple antiviral drugs. Because viruses don't have plasmids, by what mechanism is resistance arising in HIV?

Concept Building Questions

1. In Chapter 6, we learned about the growth and nutritional requirements of microbes. In particular, we learned about the growth curve of an organism in a laboratory culture (lag, log, stationary, and death phases). A patient has a bacterial infection. Describe (in words or pictures) what the growth curve

of an infection might look like inside a host. If a person is prescribed an antibiotic, what will happen to the growth curve? (Be sure to incorporate the idea of sensitive and resistant cells in your answer.) For the antibiotic growth curve, explain how the curve would look if the patient successfully completed the prescribed drug therapy, as well as what it would look like if the patient stopped taking the drug too early.

2. Explain why the development of a drug-resistant bacterial strain requires prolonged exposure to a given drug.

11 Characterizing and Classifying Prokaryotes

CHAPTER SUMMARY

General Characteristics of Prokaryotic Organisms (pp. 319–322)

Prokaryotes are by far the most diverse group of cellular microbes. This chapter examines general prokaryotic characteristics and concludes with a survey of specific prokaryotic taxa.

Morphology of Prokaryotic Cells

There are three basic shapes of prokaryotic cells: **cocci**, which are roughly spherical; **bacilli**, which are rod-shaped; and **spirals**. Spiral-shaped prokaryotes are either **spirilla**, which are stiff, or **spirochetes**, which are flexible. Slightly curved rods are **vibrios**, and the term **coccobacillus** is used to describe cells that are intermediate in shape between cocci and bacilli. In addition, there are star-shaped, triangular, and rectangular prokaryotes, as well as **pleomorphic** prokaryotes, which vary in shape and size.

Reproduction of Prokaryotic Cells

All prokaryotes reproduce asexually. The most common method of reproduction is **binary fission**, in which the parent cell disappears with the formation of progeny. A variation of binary fission called **snapping division** occurs in some Gram-positive bacilli. In snapping division, the parent cell's outer wall tears apart with a snapping movement to create the daughter cells. Other microbes reproduce asexually by different methods. For example, *actinomycetes* produce reproductive cells called **spores** that can develop into clones of the original organism; *cyanobacteria* reproduce by fragmentation; and still other prokaryotes reproduce by **budding**, in which an outgrowth of the original cell receives a copy of the genetic material, enlarges, and is then cut off from the parent cell. *Epulopiscium* and many of its relatives have a unique method of reproduction whereby the organism gives "birth" to live offspring that emerge from the body of a dead mother cell.

Arrangements of Prokaryotic Cells

Cocci are often found arranged in groups. Pairs of cocci are **diplococci**, whereas **streptococci** are long chains. Clusters are **staphylococci**, and cuboidal packets are **sarcinae**.

Bacilli are found singly, in pairs, in chains, or in a folded **palisade** arrangement.

Endospores

The Gram-positive bacteria *Bacillus* and *Clostridium* produce **endospores**. Endospores are stable resting stages that barely metabolize. A vegetative cell normally transforms itself into an endospore under hostile or unfavorable conditions.

The process is called *sporulation*. Depending on the species, a cell forms an endospore either *centrally*, *subterminally* (near one end), or *terminally* (at one end).

Modern Prokaryotic Classification (pp. 322–323)

Living things are currently classified into three domains—Archaea, Bacteria, and Eukarya—based largely on genetic relatedness. The most authoritative reference in modern prokaryotic systematics is *Bergey's Manual of Systematic Bacteriology*, which classifies prokaryotes into three phyla in Archaea and 24 phyla in Bacteria. The organization of this text's survey of prokaryotes largely follows the classification scheme in *Bergey's Manual*.

Survey of Archaea (pp. 323–325)

Archaea are classified into three phyla: Crenarchaeota, Euryarchaeota, and Korarchaeota. They have unique rRNA sequences and share other common characteristics that distinguish them from bacteria:

- They lack true peptidoglycan in their cell walls.
- Cell membrane lipids have branched hydrocarbons.
- Initial amino acid sequence in their polypeptide chains is methionine.

Though most archaea live in moderate environmental conditions, the domain Archaea includes extremophiles and methanogens.

Extremophiles

Extremophiles are microbes that require extreme conditions of temperature, pH, and/or salinity to survive. Prominent among them are **thermophiles**, which thrive at temperatures over 45°C, and **hyperthermophiles**, which live at 80°C. They are found in two of the three phyla. Their DNA, membranes, and proteins do not function properly at lower temperatures. **Halophiles** depend on high concentrations of salt to keep their cell walls intact. They are found in the phylum Euryarchaeota. The halophile *Halobacterium salinarium* synthesizes purple proteins called **bacteriorhodopsins** that harvest light energy to synthesize ATP.

Methanogens

Methanogens are obligate anaerobes in the phylum Euryarchaeota that convert CO_2, H_2, and organic acids into methane gas (CH_4). These microbes constitute the largest group of archaea. Methanogens play significant roles in the environment by converting organic wastes in pond, lake, and ocean sediments into methane. Methanogens also have useful industrial applications, such as in sewage treatment.

Survey of Bacteria (pp. 326–341)

Bacteria are currently classified according to differences in 16S rRNA sequences.

Deeply Branching and Phototrophic Bacteria

The **deeply branching bacteria** are so named because their rRNA sequences and growth characteristics lead scientists to conclude that these organisms are similar to the earliest bacteria; that is, they branched off the "tree of life" at an early stage. They are autotrophic and live in hot, acidic, and anaerobic environments, often with intense exposure to sun.

Phototrophic bacteria acquire the energy needed for anabolism by absorbing light with pigments located in non-membrane-bound thylakoids called *photosynthetic lamellae*. They can be divided into the following five groups based on their pigments and their source of electrons for photosynthesis:

1. Cyanobacteria are in phylum Cyanobacteria.
2. Green sulfur bacteria are placed in phylum Chlorobi.
3. Green nonsulfur bacteria are members of phylum Chloroflexi.
4. Purple sulfur bacteria are placed in three classes of phylum Proteobacteria.
5. Purple nonsulfur bacteria are also placed in phylum Proteobacteria.

Cyanobacteria are blue-green bacteria that generally reproduce by binary fission. Some cyanobacteria reduce atmospheric N_2 to NH_3 via a process called **nitrogen fixation**. Cyanobacteria must separate in either time or space the metabolic pathways of nitrogen fixation from those of oxygenic photosynthesis because the oxygen generated during photosynthesis inhibits nitrogen fixation. Many cyanobacteria fix nitrogen in thick-walled cells called **heterocysts**.

All of the green and purple bacteria differ from plants, algae, and cyanobacteria in two ways: they use *bacteriochlorophylls* for photosynthesis instead of *chlorophyll a*, and they are *anoxygenic*; that is, they do not generate oxygen during photosynthesis. Whereas nonsulfur bacteria derive electrons for the reduction of CO_2 from organic compounds such as carbohydrates and organic acids, sulfur bacteria derive electrons from the oxidation of hydrogen sulfide to sulfur.

Low G + C Gram-Positive Bacteria

The *G + C ratio* is the percentage of all base pairs in a genome that are guanine-cytosine base pairs. Taxonomists use this ratio in classifying microbes. Bacteria with G + C ratios below 50% are considered "low G + C bacteria"; the remainder are considered "high G + C bacteria."

Low G + C bacteria are classified within the phylum **Firmicutes**, which includes the following groups:

- Clostridia are rod-shaped, obligate aerobes, many of which form endospores. The group is named for the genus *Clostridium*, which causes gangrene, tetanus, botulism, and diarrhea. Another microbe related to *Clostridium* is *Veillonella*, which is often found in dental plaque.
- **Mycoplasmas** are Gram-positive, pleomorphic, facultative anaerobes and obligate anaerobes that lack cell walls. However, they are able to survive because they colonize osmotically protected habitats such as human bodies, and their cytoplasmic membranes contain sterols that give them strength and rigidity. Mycoplasmas are frequently associated with pneumonia and urinary tract infections.
- Low G + C Gram-positive bacilli and cocci are important in the environment, industry, and health care. They include the genus *Bacillus*, which forms endospores and is common in soils. Some bacilli are beneficial in agriculture, but others cause anthrax and food poisoning. *Listeria* can contaminate meat and milk products, and causes bacteremia and meningitis. Bacteria in the genus *Lactobacillus* are non-spore-forming rods normally found in the mouth, gastrointestinal tract, and vagina, where they provide beneficial *microbial antagonism*. They are also used in the production of yogurt, buttermilk, and pickles. *Streptococcus* causes strep throat and other

diseases, and *Enterococcus* can infect wounds, blood, and the intestines. Finally, although *Staphylococcus* is found growing harmlessly on human skin, some strains can invade the body and cause pneumonia, toxic shock syndrome, and a variety of other serious infections.

High G + C Gram-Positive Bacteria

High G + C bacteria are classified in the phylum Actinobacteria. Members of the genus *Corynebacterium* are pleomorphic aerobes and facultative anaerobes that reproduce by snapping division. They are characterized by their stores of phosphate within inclusions called **metachromatic granules**. Members of the genus *Mycobacterium*, including species that cause tuberculosis and leprosy, grow slowly and have unique, resistant cell walls containing waxy **mycolic acids. Actinomycetes** resemble fungi in that they produce spores and form filaments; this group includes *Actinomyces*, which is normally found in human mouths, *Nocardia*, which is useful in the degradation of pollutants, and *Streptomyces*, which is used to produce important antibiotics.

Gram-Negative Proteobacteria

The phylum **Proteobacteria** constitutes the largest and most diverse group of bacteria. They are all Gram-negative and share common 16S rRNA nucleotide sequences. There are five distinct classes, designated by the first five letters of the Greek alphabet:

- **Alphaproteobacteria** are typically aerobes capable of growing at very low nutrient levels. Many have unusual attachment extensions of the cell called *prosthecae. Azospirillum* and *Rhizobium* are nitrogen fixers that are important in agriculture. Members of the genus *Nitrobacter* are **nitrifying bacteria** that oxidize nitrogen compounds to NO_3 via a process called **nitrification**. Most purple nonsulfur phototrophs are alphaproteobacteria. Pathogens in this class include *Rickettsia*, which causes typhus and Rocky Mountain spotted fever, and *Brucella*, which causes brucellosis. In industry, *Acetobacter* and *Gluconobacter* are used to synthesize acetic acid, and *Agrobacterium* is used in genetic recombination in plants.
- **Betaproteobacteria,** like alphaproteobacteria, thrive in habitats with low levels of nutrients, and some genera, such as *Nitrosomonas*, are nitrifying. Interesting genera include *Neisseria*, species of which inhabit the mucous membranes of mammals and cause such diseases as gonorrhea, meningitis, and pelvic inflammatory disease. *Bordetella* causes pertussis, and *Burkholderia* colonizes the lungs of cystic fibrosis patients. Nonpathogenic betaproteobacteria include *Thiobacillus*, which is important in recycling sulfur in the environment, and *Zoogloea* and *Sphaerotilus*, two genera that form *flocs*, slimy, tangled masses of organic matter in sewage.
- **Gammaproteobacteria** constitute the largest and most diverse class. They include purple sulfur bacteria, intracellular pathogens such as *Legionella* and *Coxiella*, and **methane oxidizers**, which digest most of the methane produced by methanogens before it can adversely affect the world's climate. The largest group is composed of Gram-negative, facultatively anaerobic rods that catabolize carbohydrates by glycolysis and the pentose phosphate pathway. The group contains numerous human pathogens, such as *Escherichia coli*. Finally, **pseudomonads** are gammaproteobacteria that utilize the Entner-Doudoroff and pentose phosphate pathways for

catabolism of glucose. They include the pathogen *Pseudomonas* and the nitrogen fixers *Azotobacter* and *Azomonas*.
- **Deltaproteobacteria** include *Desulfovibrio*, a sulfate-reducing microbe important in the sulfur cycle and corrosive to iron. *Bdellovibrio* is pathogenic to bacteria. **Myxobacteria** are soil dwelling and form stalked **fruiting bodies** containing resistant, dormant **myxospores** that can survive for a decade or more before germinating and becoming vegetative cells.
- **Epsilonproteobacteria** are Gram-negative rods, vibrios, or spirals. *Campylobacter* causes blood poisoning and inflammation of the intestinal tract, and *Helicobacter* causes ulcers.

Other Gram-Negative Bacteria

Other assorted Gram-negative bacteria are classified by *Bergey's Manual* into nine phyla. They include Chlamydias, cocci typified by the genus *Chlamydia* and responsible for neonatal blindness, pneumonia, and a sexually transmitted disease. Within a host cell, chlamydias form **initial bodies** that change into smaller **elementary bodies** released when the host cell dies. *Spirochetes* are helical bacteria that live in diverse environments. *Treponema*, the agent of syphilis, and *Borrelia*, which causes Lyme disease, are examples. **Bacteroids** include *Bacteroides*, an obligate anaerobic rod that inhabits the digestive tract, and *Cytophaga*, an aerobic rod that degrades wood and raw sewage.

KEY THEMES

Chapter 11 is the first of three chapters on the essential characteristics and general classification schemes for various microbial groups. The information in these chapters rests upon the knowledge presented in the previous 10 chapters. For prokaryotes in particular, remember the following:

> *Prokaryotes are structurally quite different from other microbes:* These differences generally allow prokaryotes to be more easily grouped and studied. For this reason prokaryotic classification schemes are more precisely defined than schemes for other microorganisms.

QUESTIONS FOR FURTHER REVIEW

Answers to these questions can be found in the answer section at the back of this study guide. Refer to the answers only after you have attempted to solve the questions on your own.

Multiple Choice

1. Cocci that divide to form grapelike clusters of cells are called
   a. streptococci. c. sarcinae.
   b. staphylococci. d. spirochetes.
2. Which of the following is NOT a method of reproduction in prokaryotes?
   a. binary fission c. budding
   b. snapping division d. sexual reproduction

3. Extremophiles are found among
   a. the Crenarchaeota.
   b. the Cyanobacteria.
   c. the Firmicutes.
   d. the Actinobacteria.

4. The largest known group of Archaea are the
   a. thermophiles.
   b. hyperthermophiles.
   c. halophiles.
   d. methanogens.

5. The term *deeply branching bacteria* refers to
   a. organisms with branched flagella.
   b. organisms that "branched" early off of the tree of life.
   c. organisms with branched fatty acids in their membranes.
   d. none of the above.

6. Unlike cyanobacteria, the green and purple bacteria
   a. use chlorophyll *a* for photosynthesis.
   b. are not photosynthetic.
   c. are anoxygenic.
   d. none of the above.

7. Which of the following statements is true regarding mycoplasmas?
   a. They are fungi.
   b. They were identified as Gram-positive based on their thick cell walls.
   c. They contain sterols in their membranes.
   d. Morphologically they appear as bacilli.

8. Which of the following is NOT classified as a low G + C Gram-positive organism?
   a. *Clostridium*
   b. *Bacillus*
   c. *Streptococcus*
   d. *Corynebacterium*

9. Which of the following could be used to clearly distinguish mycoplasmas from mycobacteria?
   a. G + C content
   b. Gram stain
   c. growth characteristics on agar plates
   d. both a and c

10. Prosthecae are produced by
    a. mycoplasmas.
    b. mycobacteria.
    c. alphaproteobacteria.
    d. actinomycetes.

11. Which group of proteobacteria thrive in habitats where nutrients are in poor supply?
    a. alphaproteobacteria
    b. betaproteobacteria
    c. gammaproteobacteria
    d. both a and b

12. Which group of proteobacteria make up the largest and most diverse class?
    a. alphaproteobacteria
    b. betaproteobacteria
    c. gammaproteobacteria
    d. deltaproteobacteria

13. *E. coli*, the most common laboratory organism in use today, is an example of
    a. alphaproteobacteria.
    b. betaproteobacteria.
    c. gammaproteobacteria.
    d. deltaproteobacteria.

14. Which of the bacteria below produce fruiting bodies when nutrients are depleted?
    a. mycobacteria
    b. myxobacteria
    c. mycoplasma
    d. actinomycetes

15. Which of the following are considered gliding bacteria along with *Cytophaga*?
    a. myxobacteria
    b. mycobacteria
    c. cyanobacteria
    d. both a and c

Fill in the Blanks

1. The three basic morphologies seen among prokaryotes are ________________, ________________, and ________________.

2. In binary fission, the parent cell ________________ (remains/disappears) with the formation of progeny. The reproductive process of budding is the direct opposite, in which the parent cell ________________ (remains/disappears).

3. The three phyla of Archaea as determined by rRNA sequences are ________________, ________________, and ________________.

4. ________________ inhabit extremely saline habitats. These organisms are found in phylum ________________.

5. The proper, accepted name for blue-green algae is ________________.

6. ________________ continue to reproduce even when refrigerated, making them a contamination threat in dairy products.

7. Corynebacteria produce phosphate inclusions called metachromatic granules

8. Strepomyces produce most of the natural antibiotics available in medicine. This genus is part of the group of high G + C Gram-positive bacteria called the Actinobacteria.

9. The largest and most diverse group of bacteria is the Proteobacteria.

10. Nitrifying alphaproteobacteria reduce NH_4^+ NH_3 to NO_3^- and are found in the genus alpha.

11. The smallest mysoplamas are bacteria that are smaller than the largest viruses. They are Gram-__________ (negative/positive) but lack peptoglycan in their cell wall.

12. *Borrelia* is a Spirochetes (name the bacterial group) that causes Lyme disease in humans.

Matching

Match the prokaryote on the left with the correct description on the right. Each letter will be used only once.

1. ___ *Aquifex*
2. ___ *Azotobacter*
3. ___ *Bacillus thuringiensis*
4. ___ *Bacteroides*
5. ___ *Chlamydia*
6. ___ *Clostridium*
7. ___ *Deinococcus*
8. ___ *Coxiella*
9. ___ *Halobacterium salinarium*
10. ___ *Methanopyrus*
11. ___ *Mycobacterium*
12. ___ *Neisseria*
13. ___ *Nocardia*
14. ___ *Pyrodictium*
15. ___ *Rhizobium*
16. ___ *Rickettsia*
17. ___ *Staphylococcus aureus*

A. Gram-positive organisms associated with endospore formation

B. Extreme organism living in deep-sea vents attached to sulfur granules

C. The most studied salt-loving microbe, it uses bacteriorhodopsins for photosynthesis

D. Member of the Euryarchaeota that grows around hydrothermal vents and produces CH_4

E. The deepest branching member of prokaryotes on the tree of life

F. The organism most likely recovered from soil contaminated with radioactive wastes

G. Human pathogen that avoids digestion by white blood cells

H. Produces Bt toxin, a natural insecticide

I. Commonly found as part of the normal human microbiota of the skin

J. Genus responsible for such well-known human diseases as tuberculosis and leprosy

K. Genus of organisms that degrade many pollutants in landfills and waterways

L. Genus of microbes that grows inside the roots of leguminous plants to fix nitrogen

M. Genus of prokaryotes that are obligate intracellular parasites and cause Rocky Mountain spotted fever, among other diseases

N. Diplococcus that causes the common STD gonorrhea

O. Free-living soil prokaryote involved in nitrogen fixation

P. Genus of prokaryotes that are intracellular parasites; this group is characterized by an infectious form called an elemental body

Q. Makes up 30% of the bacteria that can be isolated from human feces

Short-Answer Questions for Thought and Review

1. Describe the process of endospore formation.

   Vegetative cell environment conditions get bad endospore forms Sporulation

2. Why are cyanobacteria essential to maintaining the ecosystem of the Earth?

   Nitrogen Fixers NH_3

3. Briefly describe the five classes of proteobacteria, highlighting the differences among the groups.

Critical Thinking

1. Even though Archaea do not cause disease in humans, they are still important to study. Discuss two reasons why. Extreme environments;
2. Methanogens produce tons of methane, a greenhouse gas, more so than humans. Why, then, are we more concerned about human production of greenhouse gases?
3. Although low G + C ratio bacteria share similar 16S rRNA sequences and high G + C ratio bacteria share similar RNA sequences, the two groups are not similar to each other. Explain why this observation makes logical sense.
4. Ribosomal studies divide the gammaproteobacteria into several subgroups. A closer examination of these subgroups shows that they are also differentiated by what other characteristic?

Concept Building Questions

1. Endospores help some prokaryotes survive harsh conditions. Explain why, structurally, endospores physically cannot metabolize. Then, explain how structure protects the spores from such diverse conditions as boiling, toxic chemicals, and radiation.
2. Describe, in terms of what you have learned in past chapters on structure, metabolism, and genetics, why the proteobacteria comprise the largest and most diverse group of bacteria. Do they do one thing or many to earn them that title?

12 Characterizing and Classifying Eukaryotes

CHAPTER SUMMARY

General Characteristics of Eukaryotic Organisms (pp. 348–353)

Eukaryotic microbes comprise a fascinating and diverse assemblage, including species that are vital for human life and also numerous human pathogens. Among the 20 most frequent microbial causes of death worldwide, six are eukaryotic. Our discussion of the general characteristics of eukaryotes begins with a survey of the events in eukaryotic reproduction.

Reproduction of Eukaryotes

Eukaryotes have a variety of methods of asexual reproduction, including binary fission, budding, fragmentation, spore formation, and schizogony, but many reproduce sexually. Algae, fungi, and some protozoa reproduce both sexually and asexually.

Eukaryotic reproduction involves two types of division: nuclear division and cytoplasmic division.

Nuclear Division

Most of the DNA in eukaryotes is packaged with histone proteins as chromosomes in the form of chromatin fibers located within nuclei. Typically, a eukaryotic nucleus has either one or two complete copies of the chromosomal portion of a cell's genome. A nucleus with a single copy of each chromosome is called a **haploid** or $1n$ nucleus, and one with two sets of chromosomes is a **diploid** or $2n$ nucleus.

A nucleus that has replicated its DNA divides via a process called **mitosis**, which proceeds in four stages:

1. In **prophase**, the cell's DNA condenses into chromatids, and the spindle apparatus forms.
2. In **metaphase**, the chromosomes line up on a plane in the middle of the cell and attach near their centromeres to microtubules of the spindle.
3. In **anaphase**, sister chromatids separate and move to opposite poles of the spindle to form chromosomes.
4. In **telophase**, nuclear envelopes form around the daughter nuclei.

Mitosis results in two nuclei with the same ploidy as the original.

Meiosis is nuclear division of diploid cells that results in four haploid daughter nuclei. It is a necessary condition for sexual reproduction in which nuclei from two different cells fuse to form a single diploid nucleus. Meiosis occurs in two stages known as *meiosis I* and *meiosis II*. Each stage has four phases: prophase, metaphase, anaphase, and telophase. Although the precise actions within the phases of meiosis I differ from those of mitosis, overall, meiosis can be considered back-to-back mitoses without the DNA replication of interphase between them. Crossing over during meiosis I produces genetic recombinations, ensuring that the chromosomes resulting from meiosis are different from the parental chromosomes.

Cytokinesis (Cytoplasmic Division)

Cytokinesis, the division of a cell's cytoplasm, typically occurs simultaneously with telophase of mitosis, though in some algae and fungi it may be postponed or may not occur at all. In these cases, mitosis produces multinucleate cells called **coenocytes**.

Schizogony

Some protozoa such as *Plasmodium* reproduce asexually within red blood cells and liver cells via a special type of reproduction called **schizogony**, in which the cell undergoes multiple mitoses to form a multinucleate **schizont**. Only then does cytokinesis occur, simultaneously releasing numerous uninucleate daughter cells called *merozoites*.

The Classification of Eukaryotic Organisms

The classification of eukaryotic microbes is problematic and has changed frequently. Historical schemes based on similarity in morphology and chemistry have been replaced with schemes based on nucleotide sequences and ultrastructural features.

Protozoa (pp. 353–360)

Protozoa are eukaryotic, unicellular organisms that lack a cell wall.

Distribution of Protozoa

Protozoa require moist environments. They are critical members of the *plankton*, free-living, drifting organisms that form the basis of aquatic food chains. Few are pathogens.

Morphology of Protozoa

Protozoa are characterized by great morphologic diversity. Some have two nuclei, others lack mitochondria, and some have *contractile vacuoles* that pump water from cells. All free-living aquatic and pathogenic protozoa exist as a motile feeding stage called a **trophozoite**, and many have a hardy resting stage called a **cyst**, which is characterized by a thick capsule and a low metabolic rate.

Nutrition of Protozoa

Most protozoa are chemoheterotrophic, obtaining nutrients by phagocytizing bacteria, decaying organic matter, other protozoa, or the tissues of a host; a few protozoa absorb nutrients from the surrounding water.

Reproduction of Protozoa

Most protozoa reproduce asexually only, by binary fission or schizogony; a few protozoa also undergo sexual reproduction by forming **gametocytes** (gametes) that fuse with one another to form a diploid cell called a **zygote**.

Classification of Protozoa

Taxonomists continue to revise and refine the classification of protozoa based on rRNA nucleotide sequencing and features made visible by electron microscopy. One such scheme classifies protozoa into six taxa: *parabasalids, diplomonads, euglenozoa*, *alveolates*, *rhizaria*, and *amoebozoa*.

Parabasala

Parabasala species lack mitochondria, but each has a single nucleus and a *parabasal body*, which is a Golgi-like structure. A well-known opportunistic pathogenic parabasalid is *Trichomonas*, which can proliferate to cause severe inflammation of the human vagina.

Diplomonadida

Diplomonadida lack mitochondria, Golgi bodies, and peroxisomes. Diplomonads have two equal-sized nuclei and multiple flagella. An example is *Giardia*, a diarrhea-causing pathogen.

Euglenozoa

The group of **euglenozoa** called **euglenoids** is composed of photoautotrophic, unicellular microbes with chloroplasts containing light-absorbing pigments. Euglenoids store food as a unique polysaccharide called *paramylon* instead of as a starch. They have a semirigid, proteinaceous, helical *pellicle* that underlies the cytoplasmic membrane and helps maintain shape.

The group of euglenozoa called **kinetoplastids** have a single large mitochondrion that contains a unique region of DNA called a *kinetoplast*. Some, such as *Trypanosoma* and *Leishmania,* are pathogenic.

Alveolates

Alveolates have cavities called *alveoli* beneath their cell surfaces. They include **ciliates**, which have cilia and two nuclei. *Paramecium* is a well-known pond-water ciliate. The alveolates called **apicomplexans** are all pathogens of animals. The name of this group refers to the *complex* of special intracellular organelles, located at the *apices* of the infective stages of these microbes that enable them to penetrate a host cell. The group of alveolates called **dinoflagellates** are unicellular microbes that have photosynthetic pigments such as carotene. They make up a large proportion of freshwater and marine plankton, and some are responsible for **red tides**.

Rhizaria

Unicellular eukaryotes called amoebae are protozoa that move and feed by means of pseudopodia and reproduce via binary fission. Beyond these common features, amoebae exhibit little uniformity. Some taxonomists classify amoebae in two kingdoms: Rhizaria and Amoebozoa.

Amoebae in the group **Rhizaria** have threadlike pseudopodia. The marine amoebae, foraminifera, have a porous shell composed of calcium carbonate. Over 90% of known foraminifera are fossil species.

Amoebae called **Radiolaria** make up another group of Rhizaria, but they have shells composed of silica. Radiolarians live in marine water as part of the marine plankton.

Amoebozoa

Amoebozoa have lobe-shaped pseudopodia and no shells. The normally free-living amoeba *Naegleria* and *Acanthamoeba* can each cause disease of the brain in humans and animals that swim in water containing them. *Entamoeba* causes potentially fatal dysentery.

Another group of amoebozoa, the **slime molds**, was once considered fungi, but the slime molds' nucleotide sequences show that they are amoebozoa. Slime molds lack cell walls and are phagocytic in their nutrition. Slime molds can be either plasmodial slime molds or cellular slime molds. **Plasmodial slime molds** are composed of multinucleate filaments of cytoplasm. **Cellular slime molds** are composed of myxamoebae that phagocytize bacteria and yeasts.

Fungi (pp. 360–370)

Fungi are chemoheterotrophic eukaryotes with cell walls that are usually composed of a strong, flexible, nitrogenous polysaccharide called **chitin**. The study of fungi is *mycology*.

The Significance of Fungi

Fungi decompose dead organisms and recycle their nutrients. *Mycorrhizae* at the roots of about 90% of all vascular plants assist the plants in absorbing water and dissolved minerals. Humans consume edible fungi and use others in the manufacture of foods, beverages, and pharmaceuticals. Fungi are important research tools in the study of metabolism, growth, and development. About 30% of fungi produce **mycoses**, fungal diseases of plants, animals, and humans.

Morphology of Fungi

The vegetative (nonreproductive) body of a fungus is called its **thallus**. The thalli of *molds* are large and composed of long, branched, tubular filaments called **hyphae. Septate** hyphae are divided into cells by crosswalls called *septa*, whereas **aseptate** hyphae are undivided and coenocytic. The hyphae intertwine to form a tangled mass called a **mycelium**, which is typically subterranean. Mushrooms and other *fruiting bodies* of molds are only small visible extensions of a vast underground mycelium. **Dimorphic** fungi produce both yeastlike and moldlike thalli.

Nutrition of Fungi

Most fungi are **saprobes**; that is, they absorb nutrients by absorption from dead organisms. Other fungi obtain nutrients from living organisms such as soil-dwelling nematodes by using modified hyphae called **haustoria** that penetrate the host tissue to withdraw nutrients. Most fungi are aerobic, though many yeasts are facultative anaerobes that obtain energy from fermentation.

Reproduction of Fungi

Fungi reproduce asexually both by budding and via asexual spores, which are categorized according to their mode of development:

- Sporangiospores form inside a sac called a sporangium.
- Chlamydospores form with a thickened cell wall inside hyphae.
- Conidiospores are produced at the tips or sides of hyphae, but not within a sac.

Most fungi also reproduce sexually via spores. In the process, haploid cells from parental thalli fuse to form a *dikaryon*. After several hours to many years, pairs of nuclei within a dikaryon fuse; meiosis restores haploid nuclei, and the nuclei are partitioned into new spores.

Classification of Fungi

Taxonomists classify fungi into four major subgroups: division *Zygomycota*, division *Ascomycota*, division *Basidiomycota*, and *Deuteromycetes*.

Division Zygomycota

Fungi in the division **Zygomycota** are coenocytic molds called *zygomycetes*. Most are saprobes. The distinctive feature of most zygomycetes is the formation of rough-walled sexual structures called **zygosporangia** that develop from the fusion of sexually compatible hyphae. Following meiosis, one of the four meiotic nuclei survives to germinate, producing a haploid sporangium that is filled with haploid spores.

Division Ascomycota

The division **Ascomycota** contains about 32,000 known species of molds and yeasts that are characterized by the formation of haploid **ascospores** within sacs called **asci**. Asci occur in fruiting bodies called *ascocarps*. *Ascomycetes* include most of the green, red, and brown fungi that spoil food, as well as plant pathogens such as the agents of Dutch elm disease. Others are beneficial in baking and brewing, research, and pharmaceuticals.

Division Basidiomycota

In the division **Basidiomycota** are almost 22,000 species of fungi. Mushrooms and other fruiting bodies of *basidiomycetes*, called **basidiocarps**, consist of tightly woven hyphae that extend into multiple projections called *basidia*, the ends of which produce sexual **basidiospores**. Most basidiomycetes decompose cellulose and lignin in dead plants and return nutrients to the soil. Many produce toxins or hallucinatory chemicals.

Deuteromycetes

The **deuteromycetes**, formerly classified in the division Deuteromycota, are an informal grouping of fungi having no known sexual stage. Recently, the analysis of rRNA sequences has revealed that most deuteromycetes belong in the division Ascomycota, and thus modern taxonomists have abandoned Deuteromycota as a formal taxon. Most deuteromycetes are terrestrial saprobes, pathogens of plants, or pathogens of other fungi. Several are pathogenic to humans.

Lichens

Lichens are economically and environmentally important organisms composed of fungi living in partnership with photosynthetic microbes, either cyanobacteria or green algae. The fungus of a lichen reproduces by spores, which must germinate and develop into hyphae that capture an appropriate cyanobacterium or alga. Alternatively, wind, rain, and small animals disperse bits of lichen called *soredia* to new locations where they can establish a new lichen if there is suitable substrate. Lichens occur in three basic shapes: *fruticose* are either erect or hanging cylinders; *crustose* grow appressed to their substrates; and *foliose* are leaflike, with margins that grow free from the substrate. Lichens provide nitrogen, and many animals eat them. Humans use lichens in the production of food, dyes, clothing, perfumes, and medicines, as well as to monitor air quality.

Algae (pp. 370–374)

Algae are simple, eukaryotic, phototrophic organisms that, like plants, carry out oxygenic photosynthesis using chlorophyll *a*. They have sexual reproductive structures in which every cell becomes a gamete. The study of algae is called *phycology*.

Distribution of Algae

Most algae are aquatic, living in the *photic zone* of fresh, brackish, and salt bodies of water.

Morphology of Algae

Algae are either unicellular, colonial, or have simple multicellular bodies called *thalli*, which are commonly composed of branched filaments or sheets. The thalli of large algae (seaweeds) are complex, with branched *holdfasts* to anchor them to rocks, stemlike *stipes*, and leaflike *blades*. They may be buoyed in the water by gas-filled bulbs called *pneumocysts*.

Reproduction of Algae

Unicellular algae reproduce asexually or sexually. Multicellular algae typically reproduce either asexually by fragmentation or sexually by an **alternation of generations** in which a haploid thallus alternates with a diploid thallus.

Classification of Algae

The classification of algae is not settled. Historically, taxonomists have used differences in photosynthetic pigments, storage products, and cell wall composition to classify algae into several groups named for the colors of their photosynthetic pigments: *Chlorophyta*, *Rhodophyta*, *Chrysophyta*, and *Phaeophyta*.

Division Chlorophyta (Green Algae)

Chlorophyta are green algae that share numerous characteristics with plants: They have chlorophylls *a* and *b*, they store sugar and starch as food reserves, and many have cell walls composed of cellulose, whereas others have walls of protein or lack walls entirely. Their 18S rRNA sequences are comparable to plants. Most are unicellular or filamentous, and they live in freshwater.

Kingdom Rhodophyta (Red Algae)

Rhodophyta contain the pigment *phycoerythrin*, the storage molecule glycogen, and cell walls of **agar** or **carrageenan**, substances used as thickening agents for the production of microbiological media as well as of numerous foods, such as ice cream, syrups, and salad dressings.

Phaeophyta (Brown Algae)

Phaeophyta contain brown pigments called *xanthopylls*, as well as carotene and chlorophyll *a* and *c*. They use the polysaccharide *laminarin* and oils as food reserves and have cell walls composed of cellulose and *alginic acid*, which is used in numerous foods as a thickener.

Chrysophyta (Golden Algae, Yellow-Green Algae, and Diatoms)

Chrysophyta are diverse in terms of cell wall composition and pigments, but all use the polysaccharide *chrysolaminarin* as a storage product. All chrysophytes contain more orange-colored carotene than chlorophyll, which accounts for their more golden coloring.

Diatoms have cell walls made of silica, arranged in nesting halves called *frustules*. They are a major component of phytoplankton and are the major source of the world's oxygen.

Water Molds (p. 375)

Scientists once classified the microbes commonly known as *water molds* as fungi because they resemble filamentous fungi in having finely branched filaments; however, they are not true molds.

Water molds have tubular cristae in their mitochondria, cell walls of cellulose, spores with two different kinds of flagella, and diploid bodies. They are placed in the kingdom Stramenopila along with diatoms, other chrysophytes, and brown algae. Water molds decompose dead animals and return nutrients to the environment. Some are crop pathogens, such as *Phytophthora infestans*, which devastated the potato crop in Ireland in the mid-19th century, causing a famine that killed over 1 million people.

Other Eukaryotes of Microbiological Interest: Parasitic Helminths and Vectors (pp. 375–377)

Microbiologists are interested in two groups of eukaryotes that are not in fact microorganisms. The first is parasitic *helminths*, which are significant because their infective stages are usually microscopic. The second is arthropod vectors—animals that carry pathogens and have segmented bodies, hard external skeletons, and jointed legs. Disease vectors belong to two classes of arthropods. The first class, *Arachnida*, includes ticks and mites. Ticks are the most important arachnid vectors. The second class, *Insecta*, includes fleas, lice, flies, mosquitoes, and kissing bugs. Mosquitoes are the most important arthropod vectors of disease, and they carry the pathogens for diseases such as malaria, yellow fever, and viral encephalitis.

KEY THEMES

Chapter 12 is the second of our three chapters on the essential characteristics and general classification schemes for various microbial groups. Chapter 12, by combining all of the eukaryotes, covers a broader group of organisms, yet it is probably still safe to say that in terms of sheer diversity, prokaryotes win out over the eukaryotes. While studying this chapter, consider the following:

> *The different groups of eukaryotic microbes are extremely heterogeneous:* Protozoa are very different from fungi, and both are quite different from algae, water molds, and slime molds. The groups live in different places and perform different functions, yet at their core they are structurally related. Helminths and arthropod vectors are included in study among the eukaryotic microbes because of their role in disease transmission.

QUESTIONS FOR FURTHER REVIEW

Answers to these questions can be found in the answer section at the back of this study guide. Refer to the answers only after you have attempted to solve the questions on your own.

Multiple Choice

1. Which of the following is NOT a method of asexual reproduction seen in eukaryotes?
   a. binary fission
   b. budding
   c. fragmentation
   d. All are seen in eukaryotes.
2. Which of the following groups could have members with haploid genomes?
   a. animals
   b. fungi
   c. helminths
   d. plants
3. DNA replication occurs during which phase listed below?
   a. anaphase
   b. interphase
   c. metaphase
   d. prophase
4. If a haploid cell undergoes mitosis but does NOT undergo cytokinesis, the resultant cell is best described as being
   a. haploid and mononucleate.
   b. haploid but multinucleate.
   c. diploid and mononucleate.
   d. diploid and multinucleate.
5. Tetrads of chromosomes, which allow for crossing over, occur in
   a. mitosis.
   b. meiosis I.
   c. meiosis II.
   d. both a and b.
6. Which of the following is NOT a defining characteristic of protozoa?
   a. They all have eukaryotic cell structure.
   b. They are all unicellular.
   c. They all lack cell walls.
   d. They are all motile, though the type of motility differs among different groups.
7. Most protozoa obtain nutrients by
   a. chemoheterotrophism.
   b. absorption of nutrients from water.
   c. photoautotrophism.
   d. parasitisim.

8. Which of the following is NOT an alveolate?
   a. apicomplexans
   b. ciliates
   c. diatoms
   d. dinoflagellates
9. Which group of protozoans below is composed entirely of human pathogens?
   a. amoebae
   b. apicomplexans
   c. ciliates
   d. euglenoids
10. Which of the following is NOT characteristic of all amoebae?
    a. They move and feed using pseudopodia.
    b. They all reproduce by binary fission.
    c. They all lack cell walls.
    d. They all form cysts.
11. Which of the following protozoa would a botanist place among the plants?
    a. amoebae
    b. archaeozoa
    c. euglenoids
    d. kinetoplastids
12. Fungi are important in which of the following ways?
    a. They are important decomposers in the environment.
    b. They are used as food.
    c. They are important research tools.
    d. All of the above are important.
13. The thallus of a mold whose hyphae are NOT divided by crosswalls is termed
    a. septate.
    b. aseptate.
    c. coenocytic.
    d. both b and c.
14. Fruiting bodies are
    a. reproductive structures of protozoa.
    b. reproductive structures of fungi.
    c. not reproductive in nature.
    d. always formed by dimorphic fungi.
15. Fungi are traditionally categorized by
    a. type of asexual spores produced.
    b. type of sexual spores produced.
    c. thallus morphology.
    d. cell wall structure.
16. Algae are NOT a unified group of organisms, but the one thing they all have in common is
    a. morphology.
    b. biochemistry.
    c. oxygenic photosynthesis.
    d. reproductive methodologies.
17. Alternation of generations occurs in
    a. protozoa.
    b. fungi.
    c. algae.
    d. water molds.
18. Which group of algae is most closely related to green plants?
    a. Chlorophyta
    b. Chrysophyta
    c. Rhodophyta
    d. Phaeophyta
19. Diatoms are members of
    a. Chlorophyta.
    b. Rhodophyta.
    c. Chrysophyta.
    d. Phaeophyta.

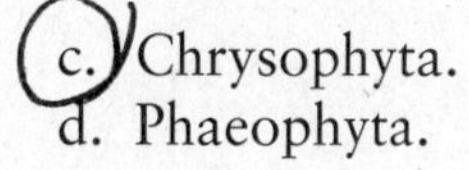

20. Which of the following statements about water molds is FALSE?
    a. They have diploid bodies.
    b. Their mitochondria have tubular cristae.
    c. They have cell walls of cellulose.
    d. All of the above are true.

Fill in the Blanks

1. ________________ and ________________ form the foundation of oceanic food chains. Additionally, however, ________________ produce toxins dangerous to humans.
2. All forms of eukaryotic reproduction involve ________________ and ________________.
3. If a diploid cell undergoes mitosis, daughter cells will be ________________ (haploid/diploid); if it undergoes meiosis, daughter cells will be ________________ (haploid/diploid).
4. Protozoa require consistently ________________ environments in which to live.
5. Many protozoa have two life stages. The feeding stage is called a(n) ________________, and the resting, or dormant, stage is called a(n) ________________.
6. Protozoans called ________________ compose large amounts of some limestone deposits.
7. The chief difference between fungi and protozoa is that fungi have ________________, whereas protozoa do not.
8. The majority of vascular plants form associations with fungi called ________________ to assist in water and mineral uptake.
9. An intertwined mass of fungal hyphae is called a(n) ________________.
10. Medical lab technicians identify fungi by the presence and type of ________________ in the sample.

11. The four main groups of fungi are Ascomycota, Basidiomycota, Zygomycota, and Deuteromycetes. Of the four, deuteromycetes do not have a known sexual cycle.

12. Red algae use the accessory pigment phycoerythrin to absorb blue light. This allows them to grow lower (higher/lower) in the water column.

13. Plasmodial slime molds are coenocytic and resemble amoebae; Cellular slime molds, in contrast, exist as individual, haploid myxamoeba.

14. Two groups of eukaryotes of interest to microbiologists even though they are not microbes are helminths and arthropod vectors.

15. The most important arthropod vector of disease is the mosquitos.

Matching

Match the eukaryote on the left with the correct description on the right. Each letter will be used only once.

1. G *Candida albicans*
2. I *Claviceps purpurea*
3. J *Gelidium*
4. E *Giardia*
5. F *Nosema*
6. B *Paramecium*
7. C *Pfiesteria*
8. K *Phytophthora infestans*
9. A *Plasmodium*
10. H *Rhizopus*
11. D *Trypanosoma*

A. Undergoes a specialized form of reproduction called schizogony
B. One of the most common ciliates found in freshwater ponds
C. Dinoflagellate that produces a potent neurotoxin that causes severe symptoms in humans
D. Pathogenic kinetoplastid that causes African sleeping sickness in humans
E. Diplomonad with two nuclei and multiple flagella
F. Microsporidian used as a biological insecticide to kill grasshoppers
G. Fungus whose growth by budding results in the formation of pseudohyphae
H. Common black bread mold
I. Ascomycete that produces LSD when it grows on grains
J. Red algae from which agar is isolated
K. Water mold responsible for the Irish potato famine of 1845

Short-Answer Questions for Thought and Review

1. List the organisms studied within the category of "eukaryotic microbes." Indicate why all the organisms listed are studied together despite their vast structural and functional differences.

2. Explain why meiosis is essential in sexually reproducing organisms.

3. What are the key distinguishing features among the six major groups of protozoa discussed in this chapter?

4. Describe the partnership of organisms that form lichens. Which types of microbes are involved, and what do they do for each other?

Critical Thinking

1. Protozoa live in moist environments, but moisture is not a constant in soil, sand, or decaying organic matter. Few protozoa are pathogens even though the internal "environment" of a human is fairly moist. Why do you suppose there aren't more protozoal pathogens?
2. Summarize the differences between fungi and plants and between fungi and animals as stated in the text. What other characteristics of fungi could be used to set them apart from the plants and animals?
3. Provide an explanation for how fungi such as *Armillaria* can reach such large sizes.
4. Compare and contrast the various ways eukaryotic microbes acquire nutrients. Which method(s) predominate? Why?

Concept Building Questions

1. Referring to previous chapters on structure, function, metabolism, and genetics, explain why the classification of fungi, algae, and protozoa is so complex. Why are taxonomists focusing primarily on nucleotide sequences and cellular ultrastructure in an attempt to reclassify many of these organisms? Explain why proper classification matters.
2. Explain metabolically why algae cannot be pathogens.

13 Characterizing and Classifying Viruses, Viroids, and Prions

CHAPTER SUMMARY

Viruses, viroids, and prions are **acellular** (noncellular) disease-causing agents that lack cell structure and cannot metabolize, grow, reproduce, or respond to their environment. To increase their numbers, they must recruit the cell's metabolic chemicals and ribosomes.

Characteristics of Viruses (pp. 383–387)

A **virus** is a minuscule, acellular, infectious agent having one or several pieces of nucleic acid—either DNA or RNA, but never both. Viruses have no cytoplasmic membrane and, with one exception, lack organelles and cytosol. In its **extracellular state**, a virus is called a **virion**. It consists of a protein coat, called a **capsid**, surrounding a nucleic acid core. Together the nucleic acid and its capsid are called a *nucleocapsid*. Some virions have a phospholipid membrane called an *envelope* surrounding the nucleocapsid. When a virus penetrates a cell, the **intracellular state** is initiated; the capsid is removed. A virus without a capsid exists solely as nucleic acid but is still referred to as a virus.

Genetic Material of Viruses

The genome of viruses includes either DNA or RNA, but never both. In addition, the DNA or RNA may be double stranded (ds) or single stranded (ss). Thus, viral genomes are described as dsDNA, ssDNA, dsRNA, or ssRNA. They may exist as multiple linear molecules of nucleic acid or circular and singular molecules of nucleic acid, depending on the type of virus. Viral genomes are usually much smaller than the genomes of cells. The smallest chlamydial bacterium has almost 1000 genes; the genome of virus MSZ has only three genes.

Hosts of Viruses

Most viruses infect only particular kinds of cells. This specificity is due to the affinity of viral surface proteins or glycoproteins for complementary proteins or glycoproteins on the surface of the host cell. A virus that infects bacteria is referred to as a **bacteriophage**, or simply a **phage**. Viruses also infect humans, other animals, plants, and even fungi.

Sizes of Viruses

Viruses are so small that only a few can be seen by light microscopy. The smallest have a diameter of 10 nm, whereas the largest are approximately 400 nm, about the size of the smallest bacterial cell.

Capsid Morphology

The capsid of a virus is composed of proteinaceous subunits called **capsomeres**. These may be composed of only a single type of protein or of several different kinds of proteins.

Viral Shapes

There are three basic types of viral shapes: Helical viruses have capsomeres that spiral around the nucleic acid, forming a tubelike structure; polyhedral viruses are roughly spherical, with a shape similar to a geodesic dome; and complex viruses have capsids of many different shapes.

The Viral Envelope

Some viruses have an **envelope** similar in composition to a cell membrane surrounding their capsids. A virus with a membrane is called an *enveloped virion*. A virion without an envelope is a *nonenveloped* or *naked virion*. The envelope of a virus is acquired from the host cell during replication or release, and it is a portion of the host cell's membrane system.

Classification of Viruses (pp. 387–389)

Virologists classify viruses by their type of nucleic acid, presence of an envelope, shape, and size. Viruses have recognized viral family and genus names. With the exception of three orders, higher taxa are not established. At this time, specific epithets for viruses are their common English designations written in italics, such as *rabies virus*.

Viral Replication (pp. 390–399)

Viruses cannot reproduce themselves because they lack the genes for all the enzymes necessary for replication, and they do not possess functional ribosomes for protein synthesis. Instead, they depend on random contact with a specific host cell type for the organelles and enzymes to produce new virions.

Lytic Replication of Bacteriophages

Viral replication that results in lysis of the cell near the end of the cycle is termed **lytic replication**. The cycle consists of five stages:

1. During **attachment**, the virion attaches to the host cell.
2. During **entry**, the virion or its genome enters the host cell. In bacteriophages, only the nucleic acid enters the cell.
3. During **synthesis**, the host cell's metabolic enzymes and ribosomes are used to synthesize new nucleic acids and viral proteins.
4. During **assembly**, new virions are spontaneously assembled in the host cell, typically as capsomeres surround replicated or transcribed nucleic acids to form new virions.
5. During **release**, new virions are released from the host cell, which lyses.

Lysogeny

Not all viruses follow the lytic pattern. Some bacteriophages have a modified replication cycle in which infected host cells grow and reproduce normally for many generations before they lyse. Such a replication cycle is called a **lysogenic replication cycle** or **lysogeny**, and the phages involved are called **lysogenic phages** or **temperate phages**. After entry into the host cell, the viral genome does not immediately assume control of the cell but instead remains inactive. Such an inactive phage is called a **prophage**. A prophage is always inserted into the DNA of the bacterium, becomes a physical part of the bacterial chromosome, and is passed on to daughter cells. Lysogenic phages can change the phenotype of a bacterium by the process of lysogenic conversion. At some point in the generations that follow, a prophage may be excised from the chromosome in a process known as **induction**. At that point, the prophage again becomes a lytic virus.

Replication of Animal Viruses

Animal viruses have the same five basic steps in their replication pathways as bacteriophages, but some differences exist, in part from the presence of envelopes around some of the viruses and in part from the eukaryotic nature of animal cells and their lack of a cell wall.

Attachment of Animal Viruses

Animal viruses lack both tails and tail fibers, and they typically attach via glycoprotein spikes or other molecules on their capsids or envelopes.

Entry and Uncoating of Animal Viruses

Some naked viruses enter their host's cells by direct penetration, a process in which the viral capsid attaches and sinks into the cytoplasmic membrane. This creates a pore through which the viral genome alone enters the cell. With other animal viruses, the entire capsid enters the cell either by membrane fusion or endocytosis. In some viruses, the viral envelope and host cell membrane fuse, releasing the capsid into the cell's cytoplasm and leaving the envelope glycoproteins as part of the cell membrane. In other cases, the entire virus is endocytized, and the capsid must be removed to release the genome. This removal of a viral capsid within a host cell is called **uncoating**.

Synthesis of Animal Viruses

Each type of animal virus requires a different strategy for synthesis that depends on the kind of nucleic acid involved: DNA or RNA, and ds versus ss:

- Synthesis of new dsDNA virions is similar to the normal replication of cellular DNA and translation of proteins. Each strand of viral DNA is used as a template for its complement. This method of replication is seen with herpesviruses and papillomaviruses; in poxviruses, synthesis occurs in the cytoplasm.
- Parvovirus, a human virus with ssDNA, is synthesized by host cell enzymes, which synthesize a complement to the ssDNA. The complementary strand binds to the ssDNA of the virus to form a dsDNA molecule. Transcription, replication, and assembly then follow.

- Some ssRNA viruses have **positive strand RNA** (+**ssRNA**), which can be directly translated by ribosomes to synthesize protein. From the +ssRNA, complementary **negative strand RNA** (−**ssRNA**) is also transcribed to serve as a template for more +ssRNA. **Retroviruses** such as HIV are +ssRNA viruses that carry *reverse transcriptase* to transcribe DNA from their RNA. This reverse process (DNA transcribed from RNA) is reflected in the name *retrovirus*.
- Viruses with −ssRNA carry an RNA-dependent RNA transcriptase for transcribing mRNA from the −ssRNA genome so that protein can then be translated. Transcription of RNA from RNA is not found in cells.
- When dsRNA functions as the genome of some viruses, one strand of the RNA molecule functions as the genome, and the other strand functions as a template for RNA replication.

Assembly and Release of Animal Viruses

Enveloped animal viruses are often released via a process called **budding**. As virions are assembled, they are extruded through one of the cell's membranes—the nuclear membrane, the endoplasmic reticulum, or the cell membrane. Each virion acquires a portion of cell membrane, which becomes its envelope. Budding allows an infected cell to remain alive for some time.

Latency of Animal Viruses

Some animal viruses, including chickenpox and herpesviruses, may remain dormant in cells in a process known as **latency**; the viruses involved are called **latent viruses** or **proviruses**. Latency may be prolonged for years. Unlike a lysogenic bacteriophage, a provirus does not typically become incorporated into the host cell's chromosomes. When it does so, as with HIV, the provirus remains there permanently; induction does not occur in eukaryotes.

The Role of Viruses in Cancer (pp. 399–400)

Neoplasia is uncontrolled cellular reproduction in a multicellular animal. A mass of neoplastic cells, called a **tumor**, may be relatively harmless (**benign**) or invasive (**malignant**). Malignant tumors are also called **cancer**. The term **metastasis** describes the spreading of malignant cells, which rob normal cells of space and nutrients, cause pain, and derange the function of affected tissues until eventually the body can no longer withstand the loss of normal function and dies.

Several theories have been proposed to explain the role viruses play in cancer. These theories revolve around the presence of *protooncogenes* that play a role in cell division. Viruses have been implicated as possible activators of oncogenes (activated protooncogenes) and as inhibitors of oncogene repressors. Among known virally induced cancers in humans are Hodgkin's disease, Kaposi's sarcoma, cervical cancer, and others.

Culturing Viruses in the Laboratory (pp. 400–402)

Because viruses cannot metabolize or replicate by themselves, they cannot be grown in standard broths or on agar plates. Instead, they must be cultured inside suitable host cells.

Culturing Viruses in Mature Organisms

Most of our knowledge of viral replication has been derived from research on bacteriophages, which are relatively easy to culture because bacteria are easily grown and maintained. Phages can be grown in bacteria maintained in either liquid cultures or on agar plates. On plates, clear zones called **plaques** are areas where phages have lysed bacteria. Such plates enable phage numbers to be estimated via a technique called **plaque assay**.

Maintaining laboratory animals can be difficult and expensive, and the practice raises ethical concerns for some. Growing viruses that infect only humans raises additional ethical complications. Therefore, scientists have developed alternative ways of culturing animal and human viruses using fertilized chicken eggs or cell cultures.

Culturing Viruses in Embryonated Chicken Eggs

Chicken eggs are a useful culture medium for viruses because they are inexpensive, are among the largest of cells, are free of contaminating microbes, and contain a nourishing yolk. Most suitable are eggs that have been fertilized.

Culturing Viruses in Cell (Tissue) Culture

Viruses can also be grown in **cell culture**, which consists of cells isolated from an organism and grown on the surface of a medium or in broth. They are of two types. **Diploid cell cultures** are created from embryonic animal, plant, or human cells that have been isolated and provided appropriate growth conditions. **Continuous cell cultures** are longer lasting because they are derived from tumor cells, which divide relentlessly.

Are Viruses Alive? (p. 402)

The characteristics of life are growth, self-reproduction, responsiveness, and the ability to metabolize. According to these criteria, viruses seem to lack the qualities of living things. For some scientists, however, three observations indicate that viruses are the least complex living entities: First, viruses use sophisticated methods to invade cells. Second, they take control of host cells. Third, they possess genomes containing the instructions for their own replication. Thus, viruses teeter on the threshold of life: Outside cells, they do not appear to be alive, but within cells, they direct the synthesis and assembly required to make copies of themselves.

Other Parasitic Particles: Viroids and Prions (pp. 402–405)

Two molecular particles also infect cells: viroids and prions.

Characteristics of Viroids

Viroids are small, circular pieces of RNA with no capsid that infect and cause disease in plants. Similar RNA molecules affect some fungi.

Characteristics of Prions

Prions are infectious protein particles that lack nucleic acids and replicate by converting similar normal proteins into new prions. Diseases associated with prions are bovine spongiform encephalopathy (BSE, so-called "mad cow disease"), scrapie in sheep, and variant Creutzfeldt-Jakob disease in humans. All of these involve fatal neurological degeneration. Normal cooking or sterilization procedures do not deactivate prions, though they are destroyed by incineration. There is no treatment for any prion disease.

KEY THEMES

Chapter 13 is the last of the three chapters on essential characteristics and general classification schemes for the various microbial groups, and it is the least precise. Viruses, viroids, and prions by their very nature defy classification in the tradition sense of biology. Remember the following while studying this chapter:

> *Viruses, viroids, and prions are acellular and do not form branches of the tree of life:* It is doubtful whether any of these microscopic entities are living, and any relationships they may have had with living cells are so remote as to be impossible to uncover. When studying viruses, viroids, and prions, we are therefore studying microbes completely different from any of the other microbes we have come to know.

QUESTIONS FOR FURTHER REVIEW

Answers to these questions can be found in the answer section at the back of this study guide. Refer to the answers only after you have attempted to solve the questions on your own.

Multiple Choice

1. Which of the following is a true statement about viruses?
   a. Viruses are self-replicating.
   b. Viruses are active outside cells.
   c. Viruses contain either DNA or RNA, but not both.
   d. Viruses are capable of growth.
2. Which of the following genome types is NOT found among viruses?
   a. dsDNA
   b. dsRNA
   c. ssRNA
   d. All of these are found among viruses.
3. Phages are viruses that specifically infect
   a. algae.
   b. bacteria.
   c. fungi.
   d. protozoa.
4. Naked virions lack
   a. a genome.
   b. a capsid.
   c. an envelope.
   d. all of the above.
5. Which step of the viral replication cycle of T4 phage is performed for the virus by *E. coli*?
   a. attachment
   b. penetration
   c. assembly
   d. All are performed by *E. coli*.
6. Which event below generally occurs as a discrete step in the replication cycle of animal viruses but NOT in that of phages?
   a. attachment
   b. penetration
   c. uncoating
   d. assembly

7. Which method of penetration can be done by both phages and naked animal viruses?
   a. direct injection of the genome across the cell membrane
   b. fusion of the capsid with the cell membrane
   c. stimulation of phagocytosis to bring the entire capsid inside
   d. None of the above are used by both.
8. Which type of viral genome most closely resembles eukaryotic mRNA?
   a. dsDNA
   b. +ssRNA
   c. −ssRNA
   d. dsRNA
9. Can an RNA virus ever integrate into a host cell genome as a provirus?
   a. No, only DNA viruses can integrate.
   b. Yes, any RNA virus can integrate.
   c. Yes, but only if the virus is a retrovirus and has reverse transcriptase.
   d. Proviruses never integrate into the host genome.
10. How can a virus cause cancer?
    a. by introducing an oncogene
    b. by stimulating oncogenes that are already present
    c. by interfering with tumor repression
    d. all of the above
11. When a human virus is studied in the laboratory, the best way to culture the virus would be to use
    a. bacteria.
    b. chicken eggs.
    c. mice.
    d. human cell culture.
12. Viroids differ from viruses in that viroids
    a. are larger than viruses.
    b. are always linear, whereas viruses are not.
    c. lack a capsid, whereas viruses always have a capsid.
    d. infect only animals, whereas viruses can infect any cell type.
13. Prions are
    a. viruses.
    b. viroids.
    c. infectious proteins.
    d. none of the above.
14. Which of the following mechanisms is NOT a method by which prions are transmitted?
    a. ingestion of infected tissue
    b. transplantion of infected tissue
    c. mucous membrane contact with infected tissue
    d. All of the above are methods of transmission.

Fill in the Blanks

1. A virion is the ____________________ (intracellular/extracellular) state of a virus. The virion consists at a minimum of a protein capsid and a nucleic acid sequence.
2. Viral capsids are constructed from protein subunits called capsomeres.

3. The three capsid types seen among viruses are helical, polyhedral, and complex.

4. Phage T4 releases an enzyme called lysozeme to aid in penetration of the host cell. This enzyme is also used during release (name the stage of replication).

5. Phages can undergo one of two types of replication inside a host cell, lytic or lysogeny.

6. Though they are dsDNA viruses, poxvirus & hep B do not replicate in the nucleus as do other dsDNA viruses.

7. parvovirus are single-stranded DNA viruses.

8. HIV is a(n) retrovirus (name the type of virus) that has a __________ genome. It carries the enzyme reverse transcriptase so it can make a DNA intermediate from its genome.

9. Enveloped viruses are released from a host cell by budding. In this process, the host cell remains alive (dies/remains alive).

10. Cells that divide uncontrollably are said to be neoplasia; a mass of such cells forms a(n) tumor. If these cells are malignant, they spread in a process called metastasis to cause cancer.

11. The two types of cell cultures are __________ and __________.

Short-Answer Questions for Thought and Review

1. List the steps in lytic phage replication. For each step, explain in one sentence what happens to the virus.

2. Summarize the key differences between a lytic replication cycle and a lysogenic replication cycle.

3. List the differences between phage replication and the replication cycle of an enveloped animal virus.

4. State the two major differences between latency and lysogeny.

Critical Thinking

1. Why would a dichotomous key be useless for helping a laboratory technician to identify viruses in a patient sample?
2. Assume an individual has the misfortune to be infected with both a bacterial pathogen and a virus at the same time. Assuming the same number of each agent was introduced and that both can actually establish infection in the human host, which microbe will be the most likely cause of disease? Why?
3. If an enveloped virus loses its envelope before it can get to a new host cell, will it be able to infect that cell? Answer yes or no, and explain.
4. For the genome types listed in Table 13.3, rank them in terms of which viruses would have the least complications or problems in replicating inside a host cell. The first virus on your list should have the easiest time replicating, and the last virus on your list should have the hardest.
5. There is some debate as to whether or not viruses are alive. Viroids and prions, however, leave little doubt as to the fact that they are not alive. Why is this? (Hint: Look at the comparisons outlined in Table 13.5.)

Concept Building Questions

1. Based on the concepts of protein structure, function, and chemical bonding discussed in previous chapters, explain how some viruses, such as HIV, have very specific host cell requirements whereas other viruses, such as rabies, can infect many cell types. How is this specificity/nonspecificity achieved?
2. Based on our studies of microbial genetics, explain why it is nearly impossible to detect the presence of a lysogenic phage embedded in the host genome. Can these infections ever be "cured"? What about latent animal virus infections?

14 Infection, Infectious Diseases, and Epidemiology

CHAPTER SUMMARY

This chapter examines the relationships between pathogenic microbes and their hosts and concludes with a discussion of epidemiology.

Symbiotic Relationships Between Microbes and Their Hosts (pp. 411–415)

Symbiosis means "to live together."

Types of Symbiosis

Microbes live with their hosts in symbiotic relationships. Types of symbiosis include **mutualism,** in which both members benefit; **parasitism,** in which a parasite benefits while the host is harmed; and **commensalism,** in which one member benefits while the other is relatively unaffected. Any parasite that causes disease is called a **pathogen.**

Normal Microbiota in Hosts

Organisms called **normal microbiota** live in and on the body. Most are commensal. Some of these microbes are resident from birth to death, whereas others are transient. The womb is an *axenic* environment, free of microbes. At birth, microorganisms come into contact with the baby's body, and the first breath and food introduce others.

How Normal Microbiota Become Opportunistic Pathogens

Opportunistic pathogens cause disease when a normal microbe is introduced into an abnormal area of the body, such as when fecal matter is deposited at the entrance to the urethra; when the immune system is suppressed; and certain changes in the body, such as consumption of antibiotics, affect normal microbiotic antagonism.

Reservoirs of Infectious Diseases of Humans (pp. 415–416)

If pathogens are to enter new hosts, they must survive in some site from which they can infect new hosts. These sites where pathogens are maintained as a source of infection are called **reservoirs of infection.** They include animal hosts, human carriers, and nonliving reservoirs.

Animal Reservoirs

Diseases that spread naturally from animals to humans are called **zoonoses.** Disease may spread through direct contact with the animal or its waste products or through an arthropod **vector.**

Human Carriers

Humans may be asymptomatic **carriers** of infection. Some carriers eventually develop the disease, whereas others remain a continued source of infection without ever becoming sick. Presumably many such healthy carriers have defensive systems that protect them from illness.

Nonliving Reservoirs

Nonliving reservoirs of infection include soil, water, food, and objects.

The Movement of Microbes into Hosts: Infection (pp. 416–419)

Exposure to Microbes: Contamination and Infection

Microbial **contamination** refers to the mere presence of microbes in or on the body. Microbial contaminants include harmless resident and transient members of the normal microbiota, as well as pathogens. Successful invasion of the body by a pathogen is called an **infection.** An infection may or may not result in disease.

Portals of Entry

The entry sites of pathogens into the human body are called **portals of entry.** They include the natural openings of the skin, such as hair follicles and sweat glands, as well as abraded skin, cuts, and scrapes. Other portals of entry are the mucous membranes lining the respiratory, gastrointestinal, urinary, and reproductive tracts, as well as the conjunctiva covering the surface of the eye. In about 2% of pregnancies, the placenta becomes a portal of entry for infection of the fetus. The *parenteral route*, by which microbes are directly deposited into deep tissues via stab wounds, surgery, etc., is a means by which portals of entry can be circumvented.

The Role of Adhesion in Infection

Adhesion (or attachment) is the process by which pathogens attach to cells. It is accomplished via a variety of structures or chemical receptors called **adhesion factors** (*adhesins*). Some bacteria have lost the ability to make adhesion factors and have thereby lost their virulence. Some pathogens interact to produce a sticky biofilm that adheres to a surface and allows bacteria to thrive.

The Nature of Infectious Disease (pp. 419–426)

Disease, also known as **morbidity,** is a condition sufficiently adverse to interfere with normal functioning of the body, or any change from a state of health. Notice that a person can be infected, for example with HIV, but not yet have the disease.

Manifestations of Disease: Symptoms, Signs, and Syndromes

Diseases manifest in different ways. **Symptoms** are subjective characteristics of a disease that the patient alone can feel, such as pain. In contrast, **signs** are objectively observable and often measurable by others, such as a high temperature. A **syndrome** is a group of symptoms and signs that collectively characterizes a particular disease or abnormal condition. For example, malaise, loss of helper T cells, diarrhea, weight loss, pneumonia, toxoplasmosis, and tuberculosis characterize AIDS.

Causation of Disease: Etiology

The study of the cause of a disease is called **etiology**. In the 19th century, Pasteur, Koch, and other scientists proposed the **germ theory of disease**, which states that disease is caused by infections of pathogenic microorganisms. **Koch's postulates** are the cornerstone of infectious disease etiology. To prove that a given infectious agent causes a given disease, a scientist must satisfy all four postulates:

1. The suspected agent must be present in every case of the disease.
2. The agent must be isolated and grown in pure culture.
3. The cultural agent must cause the disease when introduced into a healthy host.
4. The same agent must be reisolated from the diseased host.

Certain circumstances can make the use of Koch's postulates difficult or even impossible.

Virulence Factors of Infectious Agents

The ability of a microorganism to cause disease is called **pathogenicity**, and the degree of pathogenicity is called **virulence**. Pathogens have a variety of traits that enable them to cause disease; collectively, these are called **virulence factors**. Virulence factors include:

- Extracellular enzymes that enable them to dissolve structural chemicals in the body
- The ability to produce **toxins** that can harm tissues, trigger damaging immune responses, or result in **toxemia**, the presence of toxins in the bloodstream
- Antiphagocytic factors, such as hyaluronic acid capsules, and chemicals that allow them to survive inside macrophages

Exotoxins are secreted by pathogens into their environment; they include enterotoxins, neurotoxins, and cytotoxins. The body protects itself with **antitoxins**, antibodies the host forms to bind with and destroy toxins. **Endotoxin**, also known as **lipid A**, is released from the cell wall of Gram-negative bacteria and can cause systemic shock and other fatal effects.

The Stages of Infectious Diseases

Following infection, a sequence of events called the **disease process** can occur. In many cases, this process consists of five stages:

1. The **incubation period** is the time between an infection and occurrence of the first symptoms.
2. The **prodromal period** is a short time of generalized, mild symptoms, such as malaise.
3. **Illness** is the most severe stage, when signs and symptoms are most evident. Fatal diseases end at this stage.
4. **Decline** is the stage during which the patient's body gradually returns to normal as the immune response and any medical treatments vanquish the pathogen.
5. During **convalescence**, the patient recovers and tissues are repaired.

A patient may be infectious during every stage of disease.

The Movement of Pathogens Out of Hosts: Portals of Exit (pp. 426–427)

Just as infections occur through portals of entry, so pathogens must leave infected patients through **portals of exit** in order to infect others. These include bodily secretions, blood, and bodily wastes, as well as some of the portals of entry discussed earlier.

Modes of Infectious Disease Transmission (pp. 427–429)

Disease can be transmitted by numerous modes, which are somewhat arbitrarily categorized into three groups: contact transmission, vehicle transmission, and vector transmission.

Contact Transmission

Direct contact transmission of infectious disease involves person-to-person spread by bodily contact. **Indirect contact transmission** occurs when pathogens are spread from one host to another via **fomites**, inanimate objects that are inadvertently used to transfer pathogens. These objects include needles, tissues, toys, coins, and so forth. **Droplet transmission** is a third type of contact transmission: Pathogens can be transmitted via droplet nuclei that exit the body during exhalation, sneezing, and coughing.

Vehicle Transmission

Vehicle transmission is the spread of pathogens via air, drinking water, and food, as well as bodily fluids being handled outside the body. **Airborne transmission** involves the spread of pathogens to the respiratory mucous membranes of a new host via an **aerosol**, a cloud of small droplets and solid particles suspended in the air. **Waterborne transmission** is important in the spread of many gastrointestinal pathogens, including those that cause giardiasis and cholera. **Fecal-oral infection** is a major source of disease in the world, as certain worms and enteroviruses shed in the feces enter the gastrointestinal system. **Foodborne transmission** involves pathogens in and on foods that are inadequately processed, undercooked, or poorly refrigerated. **Bodily fluid transmission** can occur when blood, urine, and other fluids are handled outside the body.

Vector Transmission

Vectors are animals that transmit diseases from one host to another. **Biological vectors** not only transmit pathogens but also serve as hosts for the multiplication of a pathogen during some stage of its life cycle. **Mechanical vectors** are not required as hosts by the pathogens they transmit; such vectors only passively carry pathogens to new hosts on their feet or other body parts.

Classification of Infectious Diseases (pp. 429–430)

Infectious diseases can be classified in a number of ways, each of which has its own advantages. When grouped by time course and severity, diseases may be described as acute, subacute, chronic, or latent. An **acute** disease is one which develops rapidly but lasts a relatively short time, either resolving or causing death of the host. A **chronic** disease develops slowly but is continual or recurrent.

An example is rheumatoid arthritis. **Subacute** diseases have durations and severities that lie somewhere between acute and chronic. **Latent** diseases are those in which a pathogen remains inactive for a long period of time before producing signs and symptoms. An example is herpes.

Infectious diseases are also classified as **communicable**, if they come from another infected host, or as **contagious**, if they are easily transmitted between hosts. **Noncommunicable** infectious diseases arise from outside the host or from normal microbiota. Acne and tooth decay are examples.

Epidemiology of Infectious Diseases (pp. 430–439)

Epidemiology is the study of the location, course, and transmission of diseases within populations.

Frequency of Disease

Epidemiologists track the **incidence** of a disease (that is, the number of new cases in a given area or population in a given period of time) as well as its **prevalence** (the total number of cases in a given area or population in a given period of time). The occurrence of infectious disease can also be considered in terms of a combination of frequency and geographic distribution. A disease that normally occurs continually at a relatively stable incidence within a given population or area is said to be **endemic** to that population or area. A disease is considered **sporadic** when only a few scattered cases occur. Whenever a disease occurs at a greater frequency than normal for a population or area, it is said to be **epidemic**. A **pandemic** is an epidemic occurring on more than one continent.

Epidemiological Studies

Descriptive epidemiology is the careful recording of data concerning a disease; it often includes detection of the **index case**, the first case of the disease in a given area or population. **Analytical epidemiology** seeks to determine the probable cause of a disease, its mode of transmission, and possible means of prevention. **Experimental epidemiology** involves testing a hypothesis resulting from analytical studies concerning the cause of a disease.

Hospital Epidemiology: Nosocomial Infections

Nosocomial infections and **nosocomial diseases** are acquired by patients or staff in health care facilities. They may be **exogenous** (acquired from the health care environment, such as in air-conditioning systems or on bed rails), **endogenous** (derived from normal microbiota that become opportunistic while the patient is in the health care setting), or **iatrogenic** (induced by treatment, such as with antibiotics, or medical procedures, such as surgery).

Epidemiology and Public Health

Public health organizations use epidemiological data to promulgate rules and standards for clean, **potable** water—that is, water that is fit to drink. They also work to regulate food safety standards, to prevent disease by controlling vectors and animal reservoirs, and to educate people to make healthful choices concerning the prevention of disease.

KEY THEMES

Chapter 14 begins our study of the relationships between microbes, their environment, and the organisms within that environment, such as humans. Here we begin our look at the infection process. We will cover diseases more specifically in later chapters. As you study this chapter, focus on the following:

Microbes display a variety of relationships with animals in the environment: Some of these symbiotic relationships are mutually beneficial, but others are parasitic. Pathogens are parasites, and thus pathogens are of greatest concern to medicine.

Epidemiology is concerned with the spread and control of diseases in populations: Pathogens move between hosts in a variety of ways, and epidemiology is ultimately concerned with preventing this transmission.

QUESTIONS FOR FURTHER REVIEW

Answers to these questions can be found in the answer section at the back of this study guide. Refer to the answers only after you have attempted to solve the questions on your own.

Multiple Choice

1. For which type of symbiosis is an absolute example difficult to define?
   a. mutualism
   b. commensalism
   c. parasitisim
   d. All have specific examples.
2. Pathogens are examples of what type of lifestyle?
   a. mutualism
   b. commensalism
   c. parasitism
   d. Pathogens can come from all three symbiotic associations.
3. In the human gut, *E. coli* is an example of
   a. resident microbiota.
   b. transient microbiota.
   c. normal flora.
   d. both a and c.
4. Which of the following is NOT an example of an adhesion factor?
   a. fimbriae of bacteria
   b. cellular receptor on the plasma membrane
   c. surface glycoprotein on a virus
   d. All are adhesion factors.
5. Which of the following is a *more precise* manifestation of disease and therefore more useful in diagnosis?
   a. symptoms
   b. signs
   c. pain
   d. both a and c
6. Virulence factors can include
   a. adhesion factors.
   b. mutations.
   c. toxins.
   d. both a and c.
7. Which of the following could help a microbe to escape phagocytosis?
   a. point mutations in surface proteins
   b. endotoxin
   c. exotoxin
   d. capsule

8. During which stage of the disease process is a patient likely to be infectious?
   a. incubation
   b. illness
   c. convalescence
   d. all of the above
9. If a microbe enters a patient via the respiratory tract and establishes infection, its most likely portal of exit will be
   a. skin.
   b. mucus.
   c. respiratory droplets.
   d. urine.
10. Which of the following is NOT a true reservoir for viruses?
    a. skunk
    b. dog
    c. water
    d. human
11. If a person with a cold sneezed in your face, this would be an example of what type of transmission?
    a. direct contact
    b. droplet
    c. airborne
    d. vehicle
12. If a person with a cold sneezed across the room from you, this would be an example of what type of transmission?
    a. direct contact
    b. droplet
    c. airborne
    d. vehicle
13. Hepatitis C infection of the liver is best described as
    a. an acute disease.
    b. a chronic disease.
    c. a subacute disease.
    d. a latent disease.
14. Tooth decay caused by bacteria is an example of a
    a. communicable disease.
    b. contagious disease.
    c. noncommunicable disease.
    d. latent disease.
15. If the incidence of a disease increases, prevalence will
    a. increase.
    b. decrease.
    c. stay the same.
    d. Prevalence is not dependent on incidence.
16. Which of the following is generally NOT associated with nosocomial infections?
    a. exogenous infections
    b. endogenous infections
    c. iatrogenic infections
    d. vector-borne infections
17. Which type of diseases are the most common in the United States?
    a. colds and the flu
    b. sexually transmitted diseases
    c. gastrointestinal diseases
    d. liver diseases

Fill in the Blanks

1. The three primary types of symbiosis are mutualism, commensalism, and parasitism.
2. Those areas of your body that are NOT colonized by microbes are termed axenic. The rest of your body is inhabited by microbes of two main types: resident and transient.

3. The simple presence of microbes in or on the body is called contamination; if these microbes overcome host defenses and invade the body, this is then termed a(n) ~~opportunistic~~ infection.

4. The process of microbial attachment to cells is called adhesion.

5. The ability of a microbe to cause disease is called ~~virulence~~ pathogenicity; the degree of severity of the disease that is caused is called virulence.

6. Microbes can produce one of two types of toxins: endotoxin and exotoxin. The type of toxin that is produced only by Gram-negative bacteria is a(n) endo-.

7. Diseases that are naturally spread from normal animal hosts to humans are called zoonoses.

8. A(n) fomite is an inanimate object that is inadvertently used to transfer pathogens from one host to another.

9. Biological vectors transmit pathogens and serve as hosts for the microbe; mechanical vectors serve only to transmit pathogens.

10. communicable diseases are transmitted directly or indirectly from another infected host; non-comm. diseases, however, are not transmitted between hosts.

11. The three types of epidemiological studies are descriptive, analytical, and experimental.

Short-Answer Questions for Thought and Review

1. List the factors that create opportunities for normal flora to become pathogenic. What do these factors all have in common?

2. Rank the three major portals of entry in terms of most likely exposure to least likely exposure. In other words, by which portal of entry are you most likely to become infected?

3. Define the terms *infection* and *disease*. Why aren't all infections also diseases?

4. Briefly list the exceptions to Koch's postulates.

5. Summarize the three general transmission categories for infectious diseases. Which category, among humans, probably accounts for most transmission events?

6. What factors increase the possibility of nosocomial infections? Which factor do you think is the biggest contributor to such infections?

Critical Thinking

1. Which type of symbiotic relationship is the hardest to maintain for the microbe—an essential mutualistic relationship or parasitism? Why?
2. You and seven of your closest friends spend Saturday together. In the morning, you go to the park and toss around a baseball until lunch, at which point you drive to the mall and eat at the food court. After lunch you spend some time in the mall, get some ice cream, and go back to the park, where you hang out until early evening. You then head over to a friend's house who is having a BBQ and spend most of the night. The next day, five of you are sick with cramps and diarrhea. Using the *idea* behind Koch's postulates, how would you figure out the source of illness among your group?
3. Almost all of us have had a cold at some point in our lives. Describe your last typical cold according to the five stages of the disease process. If you have been lucky enough not to be sick recently, describe a cold or other illness in a friend.
4. You are a scientist studying two zoonotic diseases. These diseases are described below. Which disease, based on its characteristics, would pose a larger threat of human infection? Why?

Disease A	*Disease B*
• Caused by a virus • Sylvatic animal reservoir is formed by mice and other small rodents • Domestic animal reservoir is formed by cats • Mode of transmission is by contact with infected hosts	• Caused by bacteria • Sylvatic animal reservoir is formed by deer • There is no domestic animal reservoir • Mode of transmission is by flea bites

Concept Building Questions

1. Mutations that change adhesins over time can have either positive or negative effects on the ability of a microbe to stick to host cells. Explain why, genetically, viruses show a higher propensity for deleterious mutations than bacteria. Pull information from the chapters on chemistry, genetics, and the characteristics/classification of bacteria and viruses to support your answer.
2. You are an epidemiologist studying the occurrence of Disease X in a population. Disease X is caused by a virus and normally infects field mice. It is only rarely transmitted to humans by inhalation of feces. In humans, an upper respiratory tract infection results from exposure that is reminiscent of the flu. For the first five years of your study, only 1–5 cases of Disease X occurred each year per 100,000 individuals. The next two years showed 20 and 36 cases per 100,000, respectively. This year's data shows 220 cases already reported, and the year is only half over. Answer the following questions:
   a. In general, how would Koch's postulates be applied to ensure that the cases you have labeled "Disease X" actually are Disease X and not the flu?
   b. Identify the portal of entry and portal of exit for both the mouse and the human.
   c. Explain what is happening in the scenario presented in terms of incidence, prevalence, and endemic/sporadic/epidemic levels of disease. Propose at least one reason why you may be seeing the infection trend that is indicated by the case data.
   d. Based on previous material from the first 13 chapters of the text, explain what is happening in terms of the virus/host interaction.

15 Innate Immunity

CHAPTER SUMMARY

An Overview of the Body's Defenses (p. 446)

Because the cells and certain basic physiological processes of humans are incompatible with those of most plant and animal pathogens, humans have **resistance** to these pathogens. Nevertheless, we are confronted every day with pathogens that do cause disease in humans. It is convenient to cluster the body structures, cells, and chemicals that act against pathogens into three main lines of defense:

1. The first line of defense is nonspecific and part of our innate immunity. It is chiefly composed of external barriers to pathogens, especially the skin and mucous membranes.
2. The second line is also our innate immunity. It is internal and is composed of protective cells, bloodborne chemicals, and processes that inactivate or kill invaders.
3. The third line of defense is adaptive immunity, which will be examined in Chapter 16.

The Body's First Line of Defense (pp. 446–450)

The skin and mucous membranes present a formidable barrier to the entrance of microorganisms.

The Role of Skin in Innate Immunity

The skin is composed of an outer **epidermis**, which is in turn composed of multiple layers of tightly packed cells that constitute a barrier to most bacteria, fungi, and viruses. In addition, microorganisms attached to the epidermis are routinely sloughed off with the flakes of dead skin cells. The epidermis also contains phagocytic cells called **dendritic cells** or **Langerhans cells**. Their slender processes form an almost continuous network to intercept invaders.

The **dermis** lies beneath the epidermis and contains hair follicles, glands, blood vessels, and nerve endings. It contains tough collagen fibers that give the skin strength and flexibility, and its blood vessels deliver defensive cells and chemicals. Its sweat glands secrete perspiration, which contains salt, **antimicrobial peptides**, and **lysozyme**, an enzyme that destroys the cell wall of bacteria. Its sebaceous glands secrete **sebum**, an oily substance that keeps the skin pliable and lowers the pH of the skin surface.

The Role of Mucous Membranes and the Lacrimal Apparatus in Innate Immunity

Mucous membranes line the lumens of the respiratory, urinary, gastrointestinal, and reproductive tracts. The epithelial cells of the outermost layer are tightly packed but form only a thin layer. They play key roles in the diffusion of nutrients and oxygen and in the elimination of wastes. Like epidermal cells, epithelial cells are continually shed and then replaced via the cytokinesis of **stem cells**, generative cells capable of dividing to form daughter cells of a variety of types. Dendritic cells reside below the epithelium to phagocytize invaders. In the mucous membrane of the trachea, **goblet cells** secrete a sticky mucus that traps bacteria, and cilia sweep it upward, where it is expelled from the lungs via coughing. The lacrimal apparatus of the eye washes away pathogens in tears, which contain lysozyme to destroy bacteria.

The Role of Normal Microbiota in Innate Immunity

The skin and mucous membranes are normally home to a variety of microorganisms that protect the body by competing with potential pathogens, a situation called **microbial antagonism**. Some microbiota change the pH to favor their own growth or consume nutrients that might otherwise be available to pathogens. The presence of the normal microbiota stimulates the body's second line of defense. In addition, the normal microbiota of the intestines provide several vitamins important to human health.

Other First-Line Defenses

Antimicrobial peptides on the skin and in mucous membranes act against microorganisms. Stomach acid prevents the growth of many potential pathogens, and saliva contains lysozyme and washes microbes from the teeth.

The Body's Second Line of Defense (pp. 450–464)

The second line of defense includes cells such as phagocytes, antimicrobial chemicals such as complement and interferons, and the processes of inflammation and fever.

Defense Components of Blood

Blood is composed of **formed elements** (cells and parts of cells) within a fluid called **plasma**. Plasma is mostly water containing electrolytes, dissolved gases, nutrients, and protective proteins such as clotting factors, complement proteins, and antibodies. Plasma also contains compounds that transport and store iron. These compounds can play a defensive role by sequestering iron so that it is unavailable to microorganisms. Serum is that portion of plasma without clotting factors.

The formed elements are **erythrocytes** (red blood cells), **leukocytes** (white blood cells), and **platelets** (cell fragments involved in blood clotting). Leukocytes are divided into two groups according to their appearance in stained blood smears:

- **Granulocytes** have large granules that stain different colors. Of these, **basophils** stain blue with basic dye, **eosinophils** stain red to orange with the acidic dye eosin, and **neutrophils** stain lilac with a mixture of basic and acidic dyes. Basophils function to release histamine during inflammation, whereas eosinophils and neutrophils phagocytize pathogens. They exit capillaries by squeezing between the cells in a process called **diapedesis** or **emigration**.

- **Agranulocytes** do not appear to have granules when viewed via light microscopy; however, granules become visible with electron microscopy. They are of two types: **lymphocytes** are the smallest leukocytes and have nuclei that nearly fill the cell, whereas **monocytes** are large with slightly lobed nuclei. The latter leave the blood and mature into **macrophages**, which are the phagocytic cells of the second line of defense. **Wandering macrophages** perform their scavenger function while traveling throughout the body. Other macrophages are fixed and do not wander. For example, **alveolar macrophages** remain in the lungs, and **microglia** in the central nervous system. A special group of phagocytes, the dendritic cells, are located throughout the body, particularly the skin and mucous membranes.

Analysis of blood is a key tool of medical diagnosis. The proportions of leukocytes, as determined in a **differential white blood cell count**, can serve as a sign of disease.

Phagocytosis

Cells of the body capable of phagocytosis are called **phagocytes**. Phagocytosis is a continuous process that can be divided into five steps:

1. *Chemotaxis* is movement of a cell either toward or away from a chemical stimulus. **Chemotactic factors** include chemicals called **chemokines,** defensins, and other peptides derived from complement. They attract phagocytic leukocytes to the site of damage or invasion.
2. The phagocytes attach to pathogens via a process called **adherence**, in which complementary chemicals such as membrane glycoproteins bind together. Pathogens are more readily phagocytized if they are coated with antibodies or the antimicrobial proteins of complement (discussed later). This coating process is called **opsonization**, and the proteins are called **opsonins**.
3. After phagocytes adhere to pathogens, they extend pseudopodia to surround the microbe. The encompassed microbe is internalized as the pseudopodia fuse to form a food vesicle called a **phagosome**.
4. Killing occurs when lysosomes within the phagocyte fuse with newly formed phagosomes to form **phagolysosomes**, or digestive vesicles. Phagolysosomes contain antimicrobial substances in an environment with a pH of about 5.5 due to the active pumping of H^+ from the cytosol. Most pathogens are dead within 30 minutes. In the end, the phagolysosome is known as a *residual body*.
5. Phagocytes eliminate remains of microorganisms via exocytosis.

Nonphagocytic Killing

Eosinophils, **natural killer lymphocytes** (or **NK cells**), and neutrophils can accomplish killing without phagocytosis. Eosinophils attack parasitic helminths by attaching to their surface and secreting toxins and can eject mitochondrial DNA with antimicrobial activity against bacteria. *Eosinophilia*, an abnormally high number of eosinophils in the blood, is often indicative of helminth infestation. NK cells secrete toxins onto the surfaces of virally infected cells and cancerous tumors. Neutrophils can destroy nearby microbial cells without phagocytosis by creating chemicals that kill nearby invaders or by producing extracellular fibers that bind to and kill bacteria.

Nonspecific Chemical Defenses Against Pathogens

Chemicals assist phagocytic cells either by directly attacking pathogens or by enhancing other features of innate immunity. In addition to lysozyme (discussed earlier), they include Toll-like receptors, interferons, and complement.

Toll-like receptors (TLRs) are integral membrane proteins produced by phagocytic cells that recognize molecules shared by various bacterial or viral pathogens referred to as pathogen-associated molecular patterns (PAMPs). NOD proteins also recognize PAMPs, but NOD proteins are located in the cytosol rather than in the cell's membranes. Binding of a PAMP to a Toll-like receptor or NOD protein can initiate a number of defensive responses, including apoptosis of an infected cell, secretion of inflammatory mediators or interferons, or production of chemical stimulants of adaptive immune responses.

Interferons are protein molecules released by host cells to nonspecifically inhibit the spread of viral infections. They also cause the malaise, muscle aches, and fever typical of viral infections. Virally infected monocytes, macrophages, and some lymphocytes secrete **alpha interferon**, and fibroblasts secrete **beta interferon**, both of which are released within hours of infection. These interferons trigger antiviral proteins to prevent viral replication. Several days after initial infection, T lymphocytes and NK cells produce **gamma interferon**, which activates macrophages and can stimulate phagocytic activity.

Complement (the **complement system**) is a set of proteins that act as chemotactic attractants, indirectly triggers inflammation and fever, and ultimately affects the destruction of foreign cells via the formation of **membrane attack complexes (MAC)**, which puncture multiple fatal holes in pathogens' membranes. Complement is activated by a classical pathway involving antibodies, by an alternate pathway that occurs independent of antibodies, and by a lectin pathway that acts through the use of lectins.

Inflammation

Inflammation is a general, nonspecific response to tissue damage resulting from a variety of causes, including infection with pathogens. **Acute inflammation** develops quickly and is short lived. It is typically beneficial, resulting in destruction of pathogens. **Long-lasting chronic inflammation** can cause bodily damage that can lead to disease. Signs and symptoms of inflammation include redness in light-colored skin (rubor), localized heat (calor), edema (swelling), and pain (dolor).

Acute inflammation results in dilation and increased permeability of blood vessels. The process of blood clotting triggers the formation of a potent mediator of inflammation, **bradykinin**. Patrolling macrophages release other inflammatory chemicals, including **prostaglandins** and **leukotrienes**. In addition, basophils, platelets, and specialized cells located in connective tissue (**mast cells**) release histamine when they are exposed to complement fragments.

Increased blood flow delivers monocytes and neutrophils to a site of infection. As they arrive, these leukocytes roll along the inside walls of blood vessels until they adhere to the receptors lining the vessels, in a process called **margination**. They then squeeze between the cells of the vessel's wall (diapedesis) and enter the site of infection, usually within an hour of the tissue damage, to destroy the pathogens via phagocytosis.

The final stage of inflammation is *tissue repair*, which in part involves the delivery of extra nutrients and oxygen to the site. If cells called *fibroblasts* are involved in the damage, scar tissue is formed, inhibiting normal function. Cardiac muscle and certain brain tissues cannot be repaired.

Fever

Fever is a body temperature above 37°C. It results when **pyrogens**, chemicals such as interleukin 1, trigger the hypothalamic "thermostat" to reset at a higher temperature. Fever augments the beneficial effects of inflammation by enhancing the effects of interferons and inhibiting the growth of some microorganisms. It probably enhances the performance of phagocytes, the activity of cells of specific immunity, and the process of tissue repair as well. However, it has unpleasant side effects, including malaise, body aches, and tiredness; moreover, if the fever rises too high, critical proteins are denatured and nerve impulses are inhibited, resulting in seizures and even death.

KEY THEMES

Before a microbe can cause disease, it first has to adhere to a host and establish an infection. This is not an easy process. Our bodies are built to resist this invasion and we have many lines of defense to protect us. So long as these lines hold, health is maintained. When they fail or are breached, however, microbes can take over. As you read this chapter, concentrate on the following:

> *First and second lines of defense are the mechanisms by which most invaders are repelled:* The barriers of the first line of defense are the first things a microbe encounters and the first it must overcome. Second-line defenses catch the majority of what bypasses the first line. The reason we are generally as healthy as we are is that very few organisms make it past the second line.

QUESTIONS FOR FURTHER REVIEW

Answers to these questions can be found in the answer section at the back of this study guide. Refer to the answers only after you have attempted to solve the questions on your own.

Multiple Choice

1. Why is skin one of your best primary defenses?
   a. It is a tightly packed physical barrier.
   b. The outer layers shed and thus remove any adhered microbes.
   c. It is completely incapable of harboring any microbial life.
   d. Both a and b are correct.
2. Which of the following is NOT one of the first lines of defense against infection?
   a. skin
   b. mucous membranes
   c. macrophages
   d. normal flora
3. Which of the following statements is true regarding mucous membranes?
   a. Mucous membranes are as physically protective as the skin.
   b. Mucous membranes are dead, just as the outer layers of the skin are dead.
   c. The outer layers of mucous membranes are shed just as the outer layers of the skin are shed.
   d. All of the above are true.

4. Saliva contributes to first-line defense against infections by
   a. washing microbes from teeth.
   b. destroying microbes through the action of lysozyme.
   c. presenting a physical barrier to adhesion.
   d. both a and b.

5. The cells in the blood that are actively involved in defense against infection are called
   a. erythrocytes.
   b. platelets.
   c. leukocytes.
   d. goblet cells.

6. The digestive vesicles of phagocytes are formed by the fusion of
   a. antibodies and pathogens.
   b. phagosomes and lysozomes.
   c. phagosomes and lysozyme.
   d. complement and antibodies.

7. Which type of cells is known to directly and specifically kill virally infected cells and neoplastic cells?
   a. eosinophils
   b. NK cells
   c. wandering macrophages
   d. erythrocytes

8. Which organism(s) below would be most susceptible to lysis by MACs produced following initiation of the classical complement pathway? membrane attack complexes
   a. Gram-negative bacteria
   b. Gram-positive bacteria
   c. fungi
   d. viruses

9. The alternate complement pathway is NOT dependent on which of the following for functionality?
   a. C3b
   b. properdin factors
   c. microbial endotoxins or lipopolysaccharides
   d. antibodies bound to antigens

10. Which of the following is NOT a function of inflammation?
    a. increased permeability of blood vessels
    b. migration of phagocytes
    c. tissue repair
    d. All are functions of inflammation.

Fill in the Blanks

1. Skin is an __innate__ (adaptive/innate) defense against infectious agents; it is, in fact, part of your __first__ (first/second/third) line of defense.

2. Fever is an example of one of the body's __second__ lines of defense. It is an __innate__ (adaptive/innate) response to invading pathogens.

3. The three types of proteins found in blood plasma that help protect the body from infection are clotting factors, complement proteins, and antibodies.

4. Phagocytic granulocytes, specifically eosinphils and neutrophils, have the ability to leave the blood and enter tissues through a process called diapedisis or emigration.

5. Opsonization is the process whereby proteins called opsonsins are used to cover pathogens to make them more susceptible to phagocytosis.

6. Four categories of defensive chemicals used as part of second-line defenses are complement, phago, defensins, and interferons.

7. Phagocytic cells such as macrophages are capable of producing alpha interferon in response to viral infections. Furthermore, fibroblasts produce beta interferon, and T lymphocytes produce gamma interferon.

8. Fever results from the presence of chemicals called pyrogens.

Short-Answer Questions for Thought and Review

1. Explain the concept of innate resistance.
   Born w/ it, non-selective

2. Describe the main difference between innate and adaptive lines of defense.

3. Create a flowchart showing leukocyte lineages: What cells are considered leukocytes, how are they divided into groups, and what are the specific functions of each cell type?

4. Summarize the process by which inflammation allows for diapedesis of macrophages to occur.

Critical Thinking

1. Use the information in the chapter regarding the structure and function of skin and mucous membranes to explain why, for essentially healthy individuals, most respiratory tract, digestive tract, and skin infections (rashes) are rarely fatal (infections remain local and usually do not spread and cause massive, body-wide damage).
2. Other than physical structure or components, what is the primary difference between first-line defenses and second-line defenses?
3. Briefly summarize the key differences between the modes of action of complement and interferon.
4. A person is exposed to a pathogenic bacterium from a puncture wound and becomes infected. List all of the first- and second-line defenses that failed to stop the infection once the skin was broken.

Concept Building Questions

1. Describe the similarities and differences between the chemotaxis displayed by macrophages and the chemotaxis displayed by bacteria. Then, discuss how each cell can differentially respond to attractants. Bacteria, for example, respond to attractants such as nutrients, but macrophages respond to cytokines and other factors, not nutrients. Based on what you have learned in previous chapters, how does each cell type know what is in the environment and whether to respond to it or not?
2. Epidemiology is concerned with the spread of diseases in populations (and, ultimately, with controlling that spread). First- and second-line defenses influence, on an individual level, the ability to combat infections that directly affects the ability to transmit pathogens.
   a. First-line defenses are essentially barriers. If a pathogen is stopped here, what happens to the spread of a pathogen in a population?
   b. Second-line defenses are active methods of cell- and chemical-based death. What can activation of second-line defenses in individuals tell an epidemiologist about the spread of disease in a population?
   c. What does self-administration of over-the-counter medicines to suppress fever, inflammation, pain, etc., do to an epidemiologist's ability to track disease within a population?
   d. Why aren't first- and second-line defenses enough to stop the spread of diseases in a population, even if they protect us most of the time from initiation of the infection process?

16 Adaptive Immunity

CHAPTER SUMMARY

Overview of Adaptive Immunity (pp. 471–472)

Adaptive immunity is the body's ability to recognize and then mount a defense against specific invaders and their products.

Adaptive immunity has five distinct attributes:

- *Specificity.* Adaptive immune responses are precisely tailored reactions against specific attackers.
- *Inducibility.* Cells of adaptive immunity act only in response to specific pathogens.
- *Clonality.* Once induced, cells of adaptive immunity proliferate to form many generations of nearly identical cells called clones.
- *Unresponsiveness to self.* Adaptive immunity does not act against normal body cells.
- *Memory.* An adaptive immune response generates faster and more effective responses in subsequent encounters with a particular pathogen.

These aspects of adaptive immunity involve the activities of **lymphocytes,** which are a type of leukocyte (white blood cell) that acts against specific pathogens. Lymphocytes of humans form in the red bone marrow. There are two main types of lymphocytes. B lymphocytes (B cells) arise and mature in the red bone marrow of adults. T lymphocytes (T cells) begin in the bone marrow but mature in the thymus.

Adaptive immunity consists of two basic types of responses. **Humoral immune responses** involve the descendants of activated B cells and the soluble proteins they secrete called antibodies. T cells mount **cell-mediated immune responses,** which do not involve antibodies. The long-term ability of descendants of B and T cells to mount adaptive immunological responses against specific pathogens is called immunological memory.

Elements of Adaptive Immunity (pp. 472–484)

The following sections present elements of adaptive immunity by describing the "stage" and introducing the "characters" involved in adaptive immunity.

The Tissues and Organs of the Lymphatic System

The **lymphatic system** is composed of **lymphatic vessels,** which conduct the flow of *lymph*—a colorless, watery liquid similar in composition to blood plasma—from local tissues and returns it to the circulatory system. It also includes lymphoid tissues and organs that are directly involved in adaptive immunity. Lymphocytes arise in the *primary lymphoid organs* of the bone marrow and thymus. Lymphocytes then migrate to *secondary lymphoid organs and tissues*, including lymph nodes, the spleen, the tonsils, and *mucosa-associated lymphatic tissue (MALT)*.

Antigens

Antigens are portions of cells, viruses, and molecules the body recognizes as foreign. The three-dimensional shape of a region of an antigen that is recognized by the immune system is the **antigenic determinant** (or *epitope*). Effective antigen molecules are large, complex, and foreign to their host. The most effective antigens, therefore, are large foreign macromolecules such as proteins, glycoproteins, and phospholipids, but complex carbohydrates and lipids, as well as some nucleic acids, can also be antigenic.

Although immunologists characterize antigens in various ways, one important way is to group them according to their relationship to the body:

- **Exogenous antigens** include toxins and other secretions and components of microbial cell walls, membranes, flagella, and pili.
- Protozoa, fungi, bacteria, and viruses that reproduce inside a body's cells produce **endogenous antigens**. The immune system can "see" and respond to these antigens only if they are incorporated into the cell's cytoplasmic membrane.
- **Autoantigens** (or self-antigens) are antigenic molecules derived from normal cellular processes. Immune cells that treat autoantigens as foreign are normally eliminated during the development of the immune system. This phenomenon, called *self-tolerance*, prevents the body from mounting an immune response against itself.

B Lymphocytes (B Cells) and Antibodies

B lymphocytes are found in the spleen, MALT, and primarily in the germinal centers of lymph nodes. The major function of B cells is to secrete soluble antibodies. The specificity of B cell function comes from the membrane proteins called *B cell receptors*.

The surface of each B cell is covered with about 500,000 identical copies of **B cell receptor (BCR)**, a type of *immunoglobulin (Ig)*. Simple immunoglobulin contains four polypeptide chains—two *heavy chains* and two *light chains*—linked with disulfide bonds in such a way that a basic antibody molecule looks like the letter Y. A BCR has two *arms* and a transmembrane portion. The transmembrane portion anchors the BCR in the cytoplasmic membrane. The arms of each heavy and light chain vary in amino acid sequence, and thus each is called a *variable region*. The two variable regions form **antigen-binding sites**. Exact binding between an antigen-binding site and an epitope accounts for the specificity of a humoral immune response. Because an antigen typically has numerous epitopes, many different BCRs will recognize any particular antigen's epitopes, but each BCR recognizes only one epitope. When an antigenic epitope stimulates a specific B cell, B lymphocytes called **plasma cells** are produced. Plasma cells secrete immunoglobulins, called antibodies, into the blood or lymph.

Antibodies are similar to BCRs in shape but are secreted and lack the bulk of the transmembrane portion of the BCR. Antibodies carry the same specificity for an epitope as the BCR of the activated B cell. The antibody stem is called the *F_c region*. There are five basic types of F_c region and these form the five classes of antibodies: IgM, IgG, IgA, IgE, and IgD.

Once antibodies are bound to antigens, they function in several ways. These include:

- **Activation of complement** and **inflammation** (discussed in Chapter 15)
- **Neutralization** of toxins, which antibodies achieve by binding to critical portions, or neutralization of bacteria or viruses, which antibodies achieve by blocking attachment molecules on their surfaces
- **Opsonization**, or enhanced phagocytosis, whereby antibodies act as **opsonins**, molecules that stimulate phagocytosis
- **Killing by oxidation**, whereby some antibodies with catalytic properties kill bacteria directly
- **Agglutination**, which is the clumping of antibody molecules, and which may hinder the activity of pathogens and increase the chance that they will be phagocytized
- **Antibody-dependent cellular cytotoxicity** (**ADCC**), which is similar to opsonization except that the target cell dies by apoptosis

The class of antibody involved in any given humoral immune response depends on the type of invading foreign antigen, the portal of entry, and the antibody function required:

- **Immunoglobulin M** (**IgM**) is the first antibody made during the initial stages of an immune response. It is secreted as a pentamer and is most efficient at complement activation.
- **Immunoglobulin G** (**IgG**) is the most common and longest lasting class of antibody in the blood. IgG is the only antibody that can cross the placenta.
- **Immunoglobulin A** (**IgA**) is the antibody class most closely associated with various body secretions. IgA pairs with a secretory component to form **secretory IgA**.
- **Immunoglobulin E** (**IgE**) triggers mast cells and basophils to release inflammatory chemicals, such as histamine, associated with allergies.
- **Immunoglobulin D** (**IgD**) is a membrane-bound antibody molecule found in some animals as a B cell receptor.

T Lymphocytes (T Cells)

An adult's red bone marrow produces T lymphocytes. Following maturation in the thymus, T cells circulate in the lymph and blood and migrate to the lymph nodes, spleen, and Peyer's patches. Each T cell has about 500,000 copies of a **T cell receptor** (**TCR**) on its cytoplasmic membrane. A TCR does not bind epitopes directly, but instead binds only to an epitope associated with a particular protein called *MHC protein*.

Immunologists recognize different types of T cells:

- **Cytotoxic T cells** are distinguished by the presence of the CD8 cell-surface glycoprotein. They directly kill infected cells, as well as abnormal body cells such as cancer cells.
- **Helper T cells** have CD4 glycoprotein. They help to regulate the activity of B cells and cytotoxic T cells during an immune response. There are two main subpopulations: Th1 cells assist cytotoxic T cells and innate macrophages, and Th2 cells function in conjunction with B cells.
- **Regulatory T cells** express CD4 and CD25 glycoproteins. They repress adaptive immune responses and prevent autoimmune diseases.

Clonal Deletion

Cells with receptors that respond to autoantigens are selectively killed via *apoptosis* in a process known as clonal deletion (because potential offspring—clones—are deleted). Clonal deletion of T cells occurs in the thymus, where thymus cells process and present all the body's autoantigens to young T cells. Clonal deletion of B cells occurs in the bone marrow in a manner similar to that in T cells. Surviving lymphocytes and their descendants respond only to foreign antigens. When self-tolerance is impaired, the result is an *autoimmune disease.*

Immune Response Cytokines

Cytokines are soluble regulatory proteins that act as intercellular signals when released by certain body cells such as kidney cells, skin cells, and immune cells. Cytokines of the immune system include:

- **Interleukins (ILs)**, which signal among leukocytes
- **Interferons (IFNs)**, which, as discussed in Chapter 15, are antiviral proteins that also act as cytokines
- **Growth factors**, which stimulate stem cells to divide
- **Tumor necrosis factor** (TNF), which is secreted by macrophages and T cells to kill tumor cells and to regulate immune responses
- **Chemokines**, which signal leukocytes to move to a site of inflammation or infection or move within tissues

Preparation for an Adaptive Immune Response (pp. 485–487)

The body prepares for adaptive immune responses by making *major histocompatibility complex* proteins and processing antigens so that they can be recognized by T lymphocytes.

The Roles of the Major Histocompatibility Complex

The **major histocompatibility complex** is a cluster of genes located on each copy of chromosome 6 (in humans), which codes for *major histocompatibility antigens*. These antigens, which are present in the cell membranes of most cells, hold and position epitopes for presentation to T cells. Class I MHC molecules are found on the cytoplasmic membranes of all cells except red blood cells. Class II MHC proteins are found only on special cells called **antigen-presenting cells** (**APCs**). These include B lymphocytes, monocytes, macrophages, and dendritic cells. Antigens are captured, ingested, and degraded into antigenic determinants by APCs. They are then bound to MHC molecules and inserted to present the antigenic determinant on the outer surface of the APC's cytoplasmic membrane.

Antigen Processing

Antigens must be processed before MHC proteins can display epitopes. Peptides containing epitopes from endogenous antigens bind MHC class I molecules during biosynthesis in the membrane of the endoplasmic reticulum. Antigen-presenting cells internalize exogenous antigens and enzymatically catabolize the molecules to produce peptide epitopes, which bind to MHC class II molecules.

Cell-Mediated Immune Responses (pp. 487–490)

The body uses cell-mediated immune responses to fight intracellular pathogens such as viruses that have invaded body cells, as well as abnormal body cells such as cancer cells.

Activation of T Cell Clones and Their Functions

An adaptive immune response is initiated in lymphoid organs where antigen-presenting cells interact with lymphocytes. It occurs through the following series of steps:

1. Antigen presentation
2. Helper T cell differentiation
3. Clonal expansion
4. Self-stimulation

The Perforin-Granzyme Cytotoxic Pathway

In the **perforin-granzyme pathway**, cytotoxic T cells destroy their targeted cells by secreting **perforins** and **granzymes**, toxic protein molecules. Perforins perforate cell membranes, and granzymes activate apoptotic enzymes in the target cell, thereby forcing the cell to commit suicide.

The CD95 Cytotoxic Pathway

In the **CD95 cytotoxic pathway**, cytotoxic T cells contact their target cells, and their CD95L protein binds with CD95 on the target cell, thereby activating enzymes that trigger apoptosis.

Memory T Cells

Some activated T cells become **memory T cells** and can persist for months or even years in lymphoid tissues. Contact with an antigenic determinant matching its TCR prompts the memory T cell to immediately produce cytotoxic T cells. A memory response is more rapid and more effective than a primary response.

T Cell Regulation

The body carefully regulates cell-mediated immunity so that T cells do not respond to autoantigens. To prevent autoimmunity, T cells require additional signals from an APC. If they do not receive these signals, they will not respond.

Humoral Immune Responses (pp. 490–494)

Inducement of T-Independent Humoral Immunity

Because some large antigens have repeating epitopes that can be processed by B cells without the help of T cells, they are called *T-independent antigens*. The repeating subunits of T-independent antigens allow extensive cross-linking between numerous BCRs on a B cell, stimulating the B cell to proliferate. T-independent humoral immunity is relatively weak, disappears quickly, and induces little immunological memory.

Inducement of T-Dependent Humoral Immunity with Clonal Selection

T-dependent antigens lack the numerous, repetitive epitopes and large size of T-independent antigens. Immunity against them requires the assistance of type II helper T (Th2) cells. A T-dependent humoral immune response involves the following series of interactions among antigen-presenting cells and helper T cells, which are mediated by cytokines:

1. Antigen presentation for Th activation and cloning
2. Differentiation of helper T cells into Th2 cells
3. Clonal selection, in which only the B cell with BCRs complementary to the epitope will be recognized
4. Activation of B cell by the Th2 cell such that it proliferates rapidly

Memory B Cells and the Establishment of Immunological Memory

A small percentage of the cells produced by B cell proliferation do not secrete antibodies but survive as **memory B cells**, long-lived cells with BCRs complementary to the specific antigen that triggered their production. In a **primary immune response**, relatively small amounts of antibodies are produced slowly and are of limited effectiveness. When an antigen is encountered a second time, the activation of memory cells in the **secondary immune response** ensures that the immune response is rapid and strong. Enhanced immune responses to subsequent exposures are called **memory responses.**

Types of Acquired Immunity (pp. 495–496)

Immunologists categorize specific immunity as either naturally or artificially acquired, and as either active or passive.

Naturally Acquired Active Immunity

Naturally acquired active immunity occurs when the body responds to exposure to antigens by mounting specific immune responses.

Naturally Acquired Passive Immunity

Naturally acquired passive immunity occurs when a fetus, newborn, or child receives antibodies across the placenta or within breast milk.

Artificially Acquired Active Immunity

Artificially acquired active immunity occurs when the body receives antigens by injection, as with vaccinations, and mounts a specific immune response.

Artificially Acquired Passive Immunotherapy

Artificially acquired passive immunotherapy occurs when the body receives, via injection, preformed antibodies in antitoxins or antisera, which can destroy fast-acting and potentially fatal pathogens such as rabies virus.

KEY THEMES

Once past the first and second lines of defense, pathogens are faced with the last, and best, line of defense we have—the immune system. The complex association of cells, tissues, and chemicals that make up the immune system are responsible for specifically "seeing" antigens and disposing of them. As you study this chapter, focus on the following:

Active immunity is necessary for full protection against disease-causing microorganisms: Nonspecific defenses protect us against the majority of invaders, but if they are breached, the immune system has to be able to respond decisively to the challenge.

B cells are involved in tagging: One of the primary roles of B cells and antibodies is to tag foreign cells and infected cells for destruction. This is one method of controlling immune damage.

T cells are involved in killing: Once labeled for destruction, T cells clear the body of free pathogens as well as microbially infected cells.

QUESTIONS FOR FURTHER REVIEW

Answers to these questions can be found in the answer section at the back of this study guide. Refer to the answers only after you have attempted to solve the questions on your own.

Multiple Choice

1. Acquired immunity is most beneficial because it provides
   a. immediate responses to all foreign invaders.
   b. specific responses geared to specific invaders.
   c. passive barrier protection against all invaders.
   d. none of the above.

2. Which of the molecules listed below would make the best antigens?
   a. molecules with stable, defined shapes
   b. large molecules rather than small molecules
   c. visible molecules rather than hidden molecules
   d. all of the above

3. The main function of B cells is to
   a. circulate in the blood and phagocytize microbes.
   b. secrete protective antigens.
   c. secrete protective antibodies.
   d. lyse infected cells.

4. Which of the following activities of antibodies would prevent the attachment of a virus to a host cell?
   a. agglutination
   b. neutralization
   c. opsonization
   d. All of the above would prevent attachment.

5. A microbially infected cell would most likely be killed by a
   a. CD8-expressing T cell.
   b. helper T cell.
   c. cytotoxic T cell.
   d. both a and c.

6. B cells work in conjunction with which T lymphocyte?
   a. cytotoxic T cell
   b. type 1 helper T cell
   c. type 2 helper T cell
   d. B cells do not work with T cells of any type.

7. Which of the following is NOT an immune system cytokine?
   a. growth factor
   b. interferon
   c. interleukin
   d. antibody

8. Which of the following cells would possess MHC I antigens?
   a. B cell
   b. T cell
   c. skin cell
   d. all of the above

9. The best example of a T-independent antigen is a
   a. viral capsid protein.
   b. bacterial plasma membrane protein.
   c. bacterial capsule polysaccharide.
   d. fungal DNA fragment.

10. Antigen-presenting cells present antigens to
    a. type 1 helper T cells.
    b. B cells.
    c. memory cells.
    d. all of the above.

11. Which of the following statements is true about both the perforin-granzyme cytotoxic pathway and the CD95 cytotoxic pathway?
    a. Both stimulate apoptosis in the target cell.
    b. Both involve specific interaction of type 1 helper T cells with infected cells.
    c. Both involve specific interactions with cytotoxic T cells and specific receptors on the infected host cells.
    d. None of the above is true.

12. Secondary immune responses involve
    a. memory B cells.
    b. memory T cells.
    c. immediate responses without the need for activation of immune cells.
    d. all of the above.

Fill in the Blanks

1. Adaptive immunity is gained over the course of one's lifetime. This is in contrast to innate defenses, which are the same from birth.

2. Antigens are molecules that trigger specific immune responses. These molecules can further be broken down into epitopes, which are the specific, three-dimensional regions recognized by the immune system.

3. The two major types of lymphocytes are T cells and B cells.

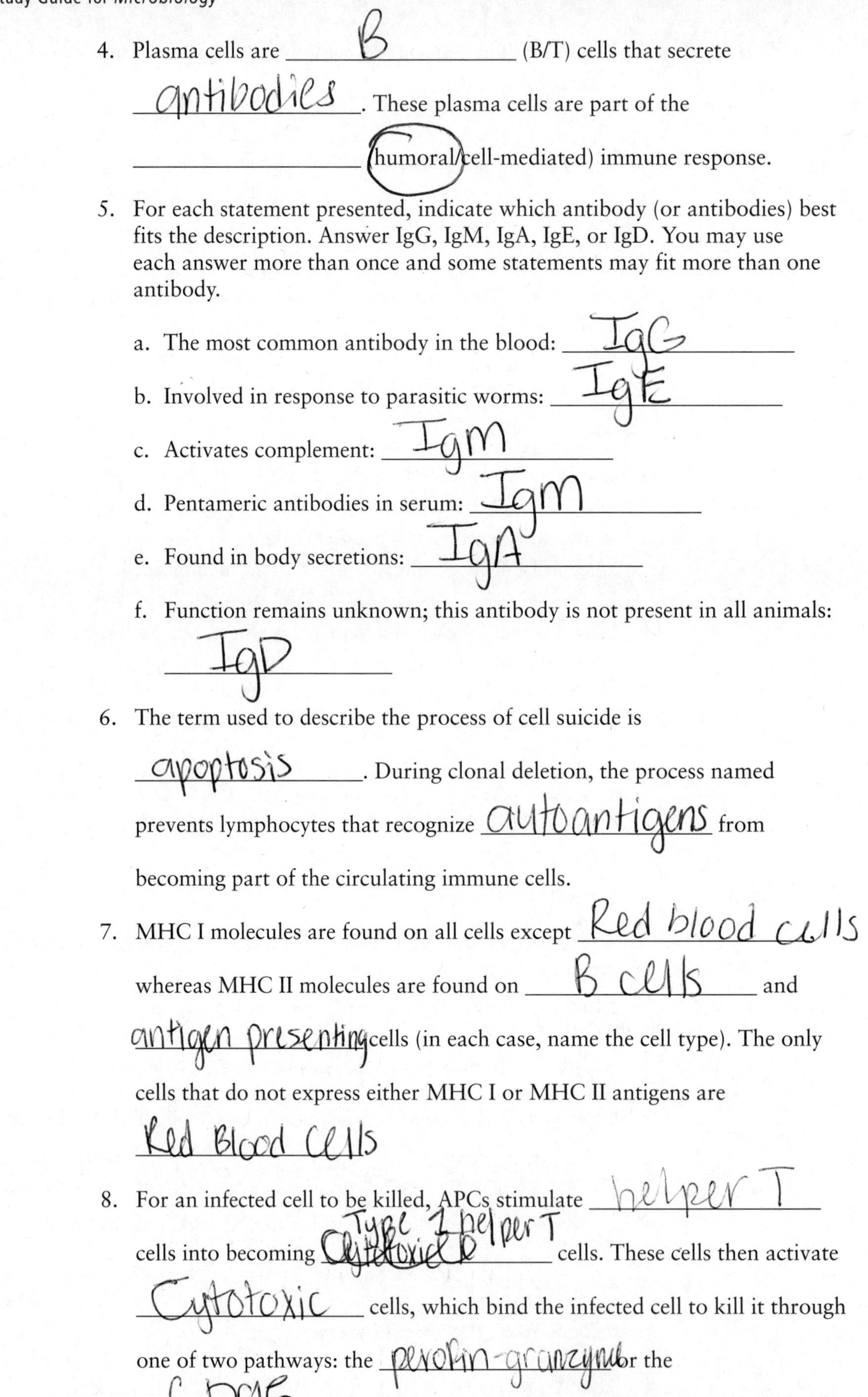

4. Plasma cells are B (B/T) cells that secrete antibodies. These plasma cells are part of the ______ (humoral/cell-mediated) immune response. [humoral circled]

5. For each statement presented, indicate which antibody (or antibodies) best fits the description. Answer IgG, IgM, IgA, IgE, or IgD. You may use each answer more than once and some statements may fit more than one antibody.

   a. The most common antibody in the blood: IgG
   b. Involved in response to parasitic worms: IgE
   c. Activates complement: IgM
   d. Pentameric antibodies in serum: IgM
   e. Found in body secretions: IgA
   f. Function remains unknown; this antibody is not present in all animals: IgD

6. The term used to describe the process of cell suicide is apoptosis. During clonal deletion, the process named prevents lymphocytes that recognize autoantigens from becoming part of the circulating immune cells.

7. MHC I molecules are found on all cells except Red blood cells whereas MHC II molecules are found on B cells and antigen presenting cells (in each case, name the cell type). The only cells that do not express either MHC I or MHC II antigens are Red Blood Cells

8. For an infected cell to be killed, APCs stimulate helper T cells into becoming ~~Cytotoxic T~~ Type 1 helper T cells. These cells then activate Cytotoxic cells, which bind the infected cell to kill it through one of two pathways: the perforin-granzyme or the CD95.

9. Acquired immune responses can be either a(n) active or passive process. Antibodies that cross the placenta from mother to child is an example of passive immunity.

10. Immune memory can only be achieved through active (active/passive) immune responses.

Short-Answer Questions for Thought and Review

1. Describe the difference between the three major types of antigens. Which type of antigen will lead to the best immune response?

2. Sketch the structure of an antibody and indicate which part designates antibody class, where antigenic determinants are bound, which parts are constant, and which parts vary.

3. Explain the absolute necessity for the process of clonal deletion.

4. Explain why clonal selection is necessary in a humoral immune response.

5. Compare and contrast a primary immune response and a secondary immune response to the same antigen.

Critical Thinking

1. Based on the structure of the lymphatic system, its tissues, vessels, and cells, explain why the immune system does not actively patrol the brain or spinal cord.
2. If you had been born with a genetic disease in which B cells are not produced, what types of immune responses would be missing?
3. Follow the fate of an exogenous pathogenic bacterium through the immune system. Assume it is phagocytized by a macrophage but that the bacterium cannot infect the macrophage.

Concept Building Questions

1. Explain why bacterial DNA is a good antigen but would not be useful in actually producing protective immunity against bacterial infection. Answer in terms of what you know about the immune system and the structure of various microbes. What other types of structures would also be antigenic but not protective?
2. Some microbes are classified partly by serotype—that is, by the antigens they display and the antibodies they elicit. Even though antigens are derived from fairly constant proteins, glycoproteins, and polysaccharides on the microbial surface, why are specific antigens a better or more useful classification parameter than the whole structural element itself?

17 Immunization and Immune Testing

CHAPTER SUMMARY

Immunization (pp. 503–511)

As we saw in Chapter 16, an individual may be made immune to an infectious disease by two artificial methods: active immunization, which involves administering antigens to a patient so that the patient actively mounts an adaptive immune response, and passive immunotherapy, in which a patient acquires immunity through the transfer of antibodies formed by immune individuals or animals. This section discusses these processes in more detail.

Brief History of Immunization

As early as the 12th century, the Chinese adopted a policy of deliberately infecting young children with particles of ground smallpox scabs from children who had survived mild cases. This procedure, called *variolation*, came into widespread use in England and in the American colonies in the early 18th century, but because it resulted in the death of 1–2% of recipients, it was outlawed. Thus, when the English physician Edward Jenner demonstrated in 1796 that protection against smallpox could be conferred by inoculation with crusts from a person infected with cowpox—a much milder disease—the technique was adopted. Because cowpox was also called *vaccinia*, Jenner called the new technique **vaccination**, and the protective inoculum a **vaccine**.

In 1879, Pasteur demonstrated that he could make an effective vaccination against *Pasteurella multocida*, and soon vaccinations against anthrax and rabies followed. After the underlying principles of immunity had been discovered, techniques of passive immunotherapy using preformed antibodies were developed. By the end of the 20th century, immunologists had developed vaccines against several potentially fatal infectious diseases, significantly reducing mortality from infectious disease worldwide. Health care providers, governments, and international organizations working together have successfully eradicated naturally occurring smallpox, and we hope for the eradication of polio, measles, mumps, and rubella. However, socioeconomic and political obstacles delay progress, and each year over 3 million children worldwide still die from vaccine-preventable infectious diseases.

Active Immunization

Vaccines include the following general types:

- **Attenuated (live) vaccines** contain attenuated microbes. The process of reducing microbial virulence is called **attenuation**, and the most common method involves raising viruses for numerous generations in culture cells to which they are not adapted, until the viruses lose the ability to produce disease. Bacteria may be made avirulent by culturing them under unusual conditions or by using genetic manipulation. Attenuated vaccines typically cause very mild infections but no true disease, and vaccinated individuals can infect those around them, thus providing immunity beyond the individual receiving the vaccine; this is called *contact immunity*. The disadvantage of attenuated vaccines is that they may retain enough virulence to cause true disease in immunosuppressed recipients, or they may revert to wild type or mutate to a virulent form.

- **Inactivated vaccines** include *whole agent vaccines*, which are produced with deactivated but whole microbes, and *subunit vaccines*, which are produced with antigenic fragments of microbes. For these vaccines, which cannot replicate or retain residual virulence, several booster doses must be administered to achieve immunity, or they must be administered with **adjuvants**, substances that increase the effective antigenicity of the vaccine. For the same reasons, vaccination with inactivated microbes does not stimulate contact immunity. However, inactivated vaccines are safer than live vaccines, although they can stimulate painful inflammation in some recipients.
- **Toxoid vaccines** are chemically or thermally modified toxins that stimulate immunity against toxins (e.g., tetanus, diphtheria) rather than cellular antigens. Effective immunity requires multiple childhood doses as well as reinoculations every 10 years of life.
- **Combination vaccines** combine antigens from several pathogens that are administered simultaneously.
- **Recombinant vaccines** are made using a variety of recombinant techniques to make vaccines less costly, more effective, and safer. These techniques have been used to produce large quantities of antigens for use in vaccines. Alternatively, an innovative method of immunization involves injecting DNA that codes for a specific antigen rather than injecting the antigen itself. Finally, recombinant technology can be used to genetically alter a microbial cell or virus so that it may be used as a live recombinant vaccine.

A common problem associated with vaccines is mild toxicity, causing pain, general malaise, and in rare cases fever high enough to promote seizures. Residual virulence (just discussed) is another potential problem. A severe problem associated with immunization is the risk of anaphylactic shock, an allergic reaction to some component of the vaccine. Because people are rarely aware of such allergies, patients who have received vaccines should remain for several minutes in the health care provider's office, where epinephrine is readily available to counter any signs of an allergic reaction.

Passive Immunotherapy

As noted earlier, passive immunotherapy involves administering preformed antibodies to a patient. Passive immunotherapy does not require the body to mount a response; instead, preformed antibodies are immediately available to bind antigen. These antibodies have in the past been obtained from people or animals that have experienced natural cases of a disease or were actively immunized against it. **Antiserum** is the blood serum (plasma without its clotting factors) of such donors. Several drawbacks (including contamination and allergic responses) of naturally harvested antiserum have led to the development of **monoclonal antibodies** developed from **hybridomas**, tumor cells created by fusing antibody-secreting plasma cells with cancerous plasma cells called *myelomas*. Hybridomas divide continuously to produce clones that secrete large amounts of a single uncontaminated antibody.

Antibody-Antigen Immune Testing (pp. 511–519)

Serology is the study and use of immunological tests to diagnose and treat disease or identify antibodies or antigens. A wide variety of serologic tests visualize antibody-antigen interactions.

Precipitation Tests

One of the simplest serologic tests relies on the fact that when antigens and antibodies are mixed in proper proportions, they form huge, insoluble, latticelike macromolecular complexes called *precipitates*. When there is excess antibody to antigen, each antigen molecule is covered with many antibody molecules, preventing extensive cross-linkage and thus precipitation. When there is excess antigen to antibody, each antigen molecule is bound to only two antibody molecules, again preventing cross-linkage; because these complexes are small and soluble, no precipitation occurs. Such small immune complexes are not easily removed and may deposit in the joints and the tiny blood vessels of the kidneys.

To ensure that optimal concentrations of antibody and antigen come together, scientists performing precipitation tests use a technique called immunodiffusion, which involves movement of the molecules through an agar gel. **Radial immunodiffusion** is a type of precipitation test in which an antigen in solution diffuses from wells into agar that contains several known concentrations of specific antibodies. The diameter of the ring of precipitate that forms around each well is compared to measure the concentrations of specific antibodies in a person's serum.

Agglutination Tests

Agglutination tests involve the clumping of particles by antibodies. In agglutination tests, technicians serially dilute the serum being tested in a process called **titration**, and then each dilution is tested for agglutinating activity. The highest dilution of serum giving a positive reaction is the **titer**. A serum that must be greatly diluted before agglutination ceases contains more antibodies than a serum that no longer agglutinates after only a few dilutions.

Neutralization Tests

Neutralization tests work because antibodies can neutralize the biological activity of many pathogens and their toxins. These are of two types:

- A **viral neutralization test** measures the ability of viruses to kill cell cultures when mixed with a patient's serum that is suspected of containing specific antibodies against them.
- A **viral hemagglutination inhibition test** can be used to detect antibodies against influenza, measles, and other viruses that naturally agglutinate red blood cells.

The Complement Fixation Test

The **complement fixation test** is a complex assay used to determine the presence of specific antibodies in an individual's serum. These tests have been replaced by other serological methods, such as ELISA or genetic analysis using polymerase chain reaction.

Labeled Antibody Tests

Labeled antibodies—those chemically linked to a fluorescent dye such as fluorescein—can be used in a variety of direct and indirect fluorescent antibody tests:

- A **direct fluorescent antibody test** allows direct observation of the presence of antigen in a tissue sample flooded with labeled antibody.
- An **indirect fluorescent antibody test** allows observation through a fluorescent microscope.
- **Enzyme-linked immunosorbent assays** (**ELISAs**) are a family of simple immune tests that use an enzyme as a label and can be readily automated and read by a machine. These tests are among the most common serologic tests used.
- The **western blot test** is a variation of an ELISA test that can detect the presence of antibodies against multiple antigens; it is used, for example, to verify the presence of antibodies against HIV in the serum of individuals who have tested positive by ELISA.

Recent Developments in Antibody-Antigen Immune Testing

Recent years have seen the development of simple immunoassays that give clinicians useful results within minutes:

- **Immunofiltration assay** is a rapid modification of the ELISA test using membrane filters rather than plates.
- **Immunochromatographic assay** is a faster and easier-to-read immunoassay in which an antigen solution flows through a porous strip, encountering labeled antibody. It is used for pregnancy testing and for rapid identification of infectious agents.

KEY THEMES

Exposure to antigens elicits immune responses. Such responses either clear the microbe from our system, allowing us to recover, or fail to clear the invader, possibly resulting in our death. Natural exposure is probably the best immune stimulus, but it is a chancy proposition, because many microbes are quite capable of killing us. Fortunately, technology has given us other options. Consider the following as you study this chapter:

Immunization allows us to protect ourselves without as much risk: No vaccine is 100% safe, but almost every vaccine is safer and less harmful than the wild-type microbe. Immunization allows us to protect ourselves individually, and as part of the herd.

Many methods exist to catalog our immune history: Various forms of immune testing help us to chart the course of disease within populations. Because few of us live completely alone all of our lives, knowing the relative health of the herd helps us to make good decisions regarding public health.

QUESTIONS FOR FURTHER REVIEW

Answers to these questions can be found in the answer section at the back of this study guide. Refer to the answers only after you have attempted to solve the questions on your own.

Multiple Choice

1. The most efficient and cost-effective means of controlling infectious disease is
   a. to allow natural infections to occur.
   b. to vaccinate people against infectious agents.
   c. to passively administer antibodies only after infection.
   d. to isolate people with contagious diseases.

2. Which infectious agent below has been completely stopped from causing natural infections thanks largely to the process of vaccination?
   a. influenza
   b. chickenpox
   c. polio
   d. smallpox

3. Attenuated vaccines can be created by
   a. propagating the microbe in cells to which it is adapted until virulence is lost.
   b. propagating the microbe in cells to which it is not adapted until virulence is lost.
   c. treating with chemicals until the microbe is inactivated.
   d. none of the above.

4. Which type of vaccine below has the least chance of reverting to a virulent form?
   a. attenuated vaccine
   b. inactivated vaccine
   c. subunit vaccine
   d. All of these can revert to virulent forms.

5. For which disease below are members of the general public usually NOT vaccinated?
   a. polio
   b. tetanus
   c. rabies
   d. hepatitis A

6. Which type of immunization would be best for an HIV patient?
   a. attenuated vaccine
   b. subunit vaccine
   c. passive vaccine
   d. None of these should be given to an HIV patient.

7. Passive immunotherapy
   a. provides rapid protection.
   b. generates an immune response.
   c. can be done once for lifelong immunity.
   d. stimulates cell-mediated immunity rather than humoral immunity.

8. In an Ouchterlony test, multiple bands between a well of antigen and a well of antibody mean that
   a. the test was done incorrectly.
   b. a single antigen and a single antibody were present, but in different amounts.
   c. several antigens and several antibodies were present.
   d. the agar plate was contaminated with unknown chemicals.

9. Agglutination tests are best performed using
   a. IgA.
   b. IgG.
   c. IgM.
   d. IgE.

10. To test serum for the presence of antibodies to a noncytopathic virus such as influenza, the best test to use would be
    a. hemagglutination.
    b. viral neutralization.
    c. immunodiffusion.
    d. viral hemagglutination inhibition.

11. To test serum for the suspected presence of very small amounts of antibody, the best test to use would be
    a. agglutination.
    b. immunoelectrophoresis.
    c. complement fixation.
    d. None of the above could be used for small amounts of antibody.
12. Labeled antibody tests do NOT include which of the tests listed below?
    a. complement fixation test
    b. fluorescent antibody test
    c. ELISA
    d. western blot test

Fill in the Blanks

1. Three general types of vaccines available are ____________________, ____________________, and ____________________ vaccines.
2. Of the types of available vaccines, ____________________ vaccines reproduce in the vaccinated individual and elicit strong immune responses.
3. ____________________ vaccines require booster shots to achieve complete immunity. ____________________ are sometimes added to increase the effective antigenicity of the vaccine.
4. ____________________ is the name given to serum that is used in passive immunotherapy. If this serum is used specifically against toxins, it is called an ____________________.
5. Immunoprecipitation tests can be done using this general method involving agar gels: ____________________.
6. In agglutination, ____________________ (soluble/insoluble) particles aggregate together, whereas in precipitation, ____________________ (soluble/insoluble) particles aggregate.
7. ____________________ describes the specific agglutination of red blood cells.
8. Direct fluorescent antibody tests are used to identify ____________________ in patient tissues; indirect fluorescent antibody tests are used to detect ____________________ in patient samples.

9. Over-the-counter pregnancy test kits are based on ____________________ assays (give the "formal" name of the assay).

Short-Answer Questions for Thought and Review

1. Give examples of how political, social, economic, and scientific issues interfere with vaccine development and distribution to the detriment of global health care.

2. Summarize the modern techniques used in the production of new vaccines. What is common to all of the techniques you listed?

Critical Thinking

1. Explain why you may not get the same cold twice, but that the second cold you get could be milder than the first.
2. You are a researcher designing a new vaccine. In clinical trials, you inoculate several people with your vaccine and need to determine not only whether an immune response was generated but also whether it would protect each subject from disease. You have no animal model to test the vaccine in, and the disease is too severe to expose humans to, simply to determine whether the vaccine works. How, then, can you determine the efficacy of your vaccine?
3. For the immunization schedule in Figure 17.3, indicate the type of microbe the vaccine is designed for, the type of vaccine used, and for which ones, as adults, you should be receiving boosters. Do you know when you were last vaccinated for the diseases in the figure? If so, indicate whether you still have protection.
4. Antibody titers are often determined to check the recovery progress of patients with infectious diseases. Explain why doing only a single antibody titer on an individual is not enough to determine whether that person is indeed recovering.
5. In general, ELISAs are fairly reliable and specific techniques. Explain why, however, they will be more reliable, specific, and accurate later in the course of disease, or even after recovery, than they will be very soon after infection.

Concept Building Questions

1. As a clinician, you encounter a patient who has recently recovered from a bacterial infection. You believe the infection was due to the same bacterium that made several of your other patients sick. This bacterium (Bacterium X) produces fairly serious disease in a majority of patients that get it, so you want to confirm your suspicions. In your most recent case you therefore need to confirm identity of the bacteria to which the patient was exposed and determine the specific class of antibodies produced against the infecting agent.

a. Design a diagnostic test using immunodiffision to identify the bacteria and the antibodies produced (Hint: You don't have to do this all in one test). What would your controls be for your tests?
b. Which class of antibodies would you expect to see if Bacterium X infects the respiratory system? What if X infected the blood (primary exposure)?
c. How could you tell the difference between an antibody that fully recognized your antigen and an antibody that only partially recognized your antigen? What is the basis of this recognition?

18 AIDS and Other Immune Disorders

CHAPTER SUMMARY

Hypersensitivities (pp. 525–538)

Immunological responses may give rise to an inflammatory reaction called a **hypersensitivity**, which may be more precisely defined as any immune response against a foreign antigen that is exaggerated beyond the norm. For example, an asthma attack prompted by a visit to a home with pet birds is an example of a hypersensitivity response.

Type I (Immediate) Hypersensitivity

Type I hypersensitivities are localized or systemic reactions that result from the release of inflammatory molecules such as histamine within seconds or minutes following contact with antigen. They are commonly called **allergies** and the antigens that stimulate them are called **allergens**.

For reasons that are not entirely clear, in allergic individuals, plasma cells produce IgE. This IgE binds very strongly with its stem to three types of defense cells—mast cells, basophils, and eosinophils—sensitizing them to respond to subsequent exposures to the allergen. Then, when the same allergen enters the body, it binds to the active sites of IgE molecules on the surfaces of sensitized cells. This binding triggers a cascade of reactions that causes the sensitized cells to release inflammatory chemicals from their granules (*degranulation*). One significant chemical released is **histamine**, a small molecule that stimulates smooth muscle contraction and vasodilation and irritates nerve endings, causing itching and pain. **Kinins** are also powerful degranulating chemicals. **Proteases** are enzymes that destroy nearby cells, activating the complement system, which results in the release of yet more inflammatory chemicals. In addition, the binding of allergens to IgE on mast cells activates other enzymes that trigger the production of **leukotrienes** and **prostaglandins**, lipid molecules that are very powerful inflammatory agents.

Depending on the amount and site at which these molecules are released, type I hypersensitivity reactions can produce various clinical syndromes. **Hay fever** is localized to the upper respiratory tract and is marked by a runny nose, sneezing, itchy throat and eyes, and excessive tear production. **Asthma** is a chronic hypersensitivity disorder affecting the lungs and characterized by constriction of the smooth muscle of the bronchi and excessive production of a thick, sticky mucus. Round, raised, itchy skin lesions characterize **urticaria**, commonly called *hives*. When the inflammatory mediators exceed the body's coping mechanisms, **acute anaphylaxis** may occur. This is a form of systemic shock in which violent constriction of the bronchial smooth muscle, widespread vasodilation, and resultant swelling of the larynx and other tissues threaten the patient's life. Without immediate administration of epinephrine, an individual in anaphylactic shock may suffocate, collapse, and die within minutes.

Clinicians diagnose type I hypersensitivity with a test that detects the amount of IgE directed against an allergen. High levels of a specific IgE indicate a hypersensitivity against that allergen.

Type I hypersensitivity is alternatively diagnosed by inoculating the skin with a very small amount of a dilute solution of the allergens being tested. Local swelling indicates sensitivity to that allergen. Avoidance of that allergen is the most effective means of prevention, but this is not always possible, as in the case of many airborne allergens. Thus, some patients elect *immunotherapy*, which consists of a series of allergy shots administered over a period of several months. However, immunotherapy must be repeated every two to three years, reduces symptoms by only about 50% in about two-thirds of patients, and is not effective in preventing asthma. For treating hay fever, antihistamines are administered. Asthma is treated with bronchodilators and various forms of corticosteroids, which are potent anti-inflammatory agents. Epinephrine quickly neutralizes anaphylaxis.

Type II (Cytotoxic) Hypersensitivity

Type II hypersensitivity reactions result when cells are destroyed by an immune response, typically by the combined activities of complement and antibodies. So-called cytotoxic hypersensitivity is part of many autoimmune diseases (discussed shortly), but the most significant examples are the destruction of donor red blood cells following an incompatible transfusion, and the destruction of fetal red blood cells.

The surface molecules of red blood cells, called **blood group antigens**, have many useful functions and vary in complexity. The ABO group system, for example, consists of just two antigens, A and B, and individuals can have either one (in which their blood type would be either A or B), both (AB blood type), or neither (blood type O). If blood is transfused to an individual with a different blood type, the donor's blood group antigens may stimulate the production of antibodies in the recipient that bind to and eventually destroy the transfused cells. The result can be a potentially life-threatening *transfusion reaction*.

Rh antigens are proteins common to the red blood cells of 85% of humans as well as rhesus monkeys. If an Rh-negative mother is pregnant with an Rh-positive fetus, fetal red blood cells may escape into the mother's blood, triggering the production of antibodies against the fetus. Initially, only IgM antibodies are produced, and because they are too large to cross the placenta, no problems arise during the first pregnancy. In additional pregnancies with an Rh-positive fetus, smaller IgG molecules directed against Rh antigen do cross the placenta, destroying fetal red blood cells and causing a variety of birth defects. Such destruction, called **hemolytic disease of the newborn**, may be limited or, especially in a third or subsequent pregnancy, severe enough to kill the baby. Fortunately, administering anti-Rh immunoglobulin (Rhogam) to Rh-negative women prevents this disease.

Drugs bound to blood platelets may subsequently bind antibodies and complement, causing immune **thrombocytopenic purpura**, in which the platelets lyse, causing hemorrhages under the skin. Similar destruction of leukocytes is one form of *agranulocytosis*, and that of red blood cells is called *hemolytic anemia*.

Type III (Immune Complex-Mediated) Hypersensitivity

As discussed in earlier chapters, the formation of complexes of antigen bound to IgG, also called **immune complexes**, initiates several molecular processes, including complement activation. Normally, immune complexes are removed from the body via phagocytosis. However, in type III hypersensitivity reactions, the immune complexes escape phagocytosis and circulate in the bloodstream until they become trapped in tissues (such as the walls of blood vessels), joints, and organs. There, they trigger mast cells to degranulate, releasing inflammatory chemicals that damage the tissues. A variety of local and systemic disorders can result. Two localized disorders include the following:

- Type III hypersensitivity reactions in the lungs cause a form of pneumonia called **hypersensitivity pneumonitis**. Individuals become sensitized when minute mold spores or other antigens are inhaled deep into the lungs, stimulating the production of antibodies. Subsequent inhalations stimulate the formation of immune complexes that then activate complement.
- **Glomerulonephritis** occurs when immune complexes circulating in the bloodstream are deposited in the walls of the glomeruli, the minute blood vessels of the kidneys. Damage to the vessels can result in kidney failure, as the glomeruli lose their ability to filter wastes from the body.

Two systemic disorders are:

- **Rheumatoid arthritis** (RA) commences when B cells secrete IgM that binds to certain IgG molecules. The IgM-IgG complexes activate complement and mast cells, which release inflammatory chemicals. The resulting inflammation causes the tissues to swell, resulting in severe pain.
- **Systemic lupus erythematosus (SLE or lupus)** is a disorder in which the individual produces autoantibodies against numerous antigens, including nucleic acids, resulting in a red rash, hemolytic anemia, bleeding disorders, and muscle inflammation.

Type IV (Delayed or Cell-Mediated) Hypersensitivity

When certain antigens contact the skin of sensitized individuals, they provoke inflammation that begins to develop at the site after 12–24 hours. Such **delayed hypersensitivity reactions** result not from the action of antibodies, but rather from interactions among antigen, antigen-presenting cells (APCs), and T cells; thus, this reaction is also called *cell-mediated hypersensitivity*. The delay reflects the time it takes for macrophages and T cells to migrate to and divide at the site of the antigen.

A good example of a delayed hypersensitivity reaction is the **tuberculin response**, generated when tuberculin, a protein extract of *Mycobacterium tuberculosis*, is injected into the skin of an individual who has been infected with or vaccinated against *M. tuberculosis*. Another example of a type IV hypersensitivity reaction is **allergic contact dermatitis**, which is T cell-mediated damage to chemically modified skin cells, such as those that have contacted poison ivy.

Graft rejection is a special case of type IV hypersensitivity. **Grafts** are tissues or organs that have been transplanted, either between sites within an individual or between a donor and a recipient. These are of several types:

- In an **autograft**, tissues are moved to a different location within the same patient.
- In an **isograft**, tissues are moved between genetically identical individuals (identical twins).
- In an **allograft**, tissues are transplanted from a donor to a genetically dissimilar recipient.
- In a **xenograft**, tissues are transplanted between individuals of different species.

Graft-versus-host disease can develop when donated bone marrow T cells recognize the recipient's cells as foreign. Physicians examine the white cells of potential graft recipients to identify a donor whose MHC proteins closely match.

The Actions of Immunosuppressive Drugs

Commonly used immunosuppressive drugs include **glucocorticoids** such as prednisone, which suppress the response of T cells to antigen and inhibit T cell cytotoxicity and cytokine production. **Cytotoxic drugs** such as cyclophosphamide and azathioprine inhibit mitosis and cytokinesis. Drugs such as cyclosporine are potent inhibitors of T cell function and can enhance the survival of allografts. Finally, *lymphocyte-depleting therapies* involve treatment with antibodies against cells of adaptive immunity.

Autoimmune Diseases (pp. 538–540)

Autoimmune diseases are any of a group of diseases that result when an individual begins to make autoantibodies or cytotoxic T cells against normal body components.

Causes of Autoimmune Diseases

There are many theories concerning the cause of autoimmune diseases, which are far more common in women than in men and tend to arise as an individual ages. These theories include the following:

- Estrogen may stimulate the destruction of tissues by cytotoxic T cells.
- Maternal cells may cross the placenta and colonize the fetus.
- Viral infections or other environmental factors may trigger autoimmune responses.
- Genetic factors may play a role, including MHC genes that in some way promote autoimmunity.
- T cells encountering self-antigens that are normally "hidden" may trigger autoimmune disease.
- Infections may trigger autoimmunity as a result of **molecular mimicry**, in which an infectious agent has an epitope that is similar to a self-antigen.
- Failure of the normal control mechanisms of the immune system may permit abnormal T cells to survive and cause disease.

Examples of Autoimmune Diseases

One group of autoimmune diseases involves only a single organ or cell type. Examples include the following:

- **Autoimmune hemolytic anemia**, a disease resulting when an individual produces antibodies against his or her own red blood cells
- **Type I (juvenile-onset) diabetes mellitus**, an immunological attack on the islets of Langerhans cells resulting in the inability to produce the hormone insulin
- **Graves' disease**, the production of autoantibodies that stimulate thyroid cells, which elicits excessive production of thyroid hormone and growth of the thyroid gland

- **Multiple sclerosis**, a disorder in which cytotoxic T cells attack and destroy the myelin sheath that insulates neurons
- **Rheumatoid arthritis**, a crippling, autoimmune disease resulting from a type III hypersensitivity reaction in which autoantibodies are formed against connective tissue in joints

Immunodeficiency Diseases (pp. 540–549)

Immunodeficiency diseases may be classified into two types:

- **Primary immunodeficiency diseases** result from mutations or developmental anomalies and occur in infants and young children
- **Acquired** or **secondary immunodeficiency diseases** result from other known causes, such as viral infections

Primary Immunodeficiency Diseases

One of the more important primary immunodeficiency diseases is **chronic granulomatous disease**, in which children have recurrent infections characterized by the inability of their phagocytes to destroy bacteria.

Some children fail to develop any lymphoid stem cells whatsoever and cannot mount any type of immune response. The resulting immune defects cause **severe combined immunodeficiency disease** (**SCID**). Other children suffer from T cell deficiencies alone. For example, **DiGeorge syndrome** results from a failure of the thymus to develop; thus, there are no T cells, and children generally die of viral infections. B cell deficiencies also occur, including **Bruton-type agammaglobulinemia**, an inherited disease in which affected babies, usually boys, cannot make immunoglobulins and experience recurrent bacterial infections. Some children may experience an inability to manufacture only one specific class of immunoglobulin.

Acquired Immunodeficiency Diseases

Previously healthy immune systems may become damaged in very old age and by severe stress, malnutrition, or environmental toxins. Insufficient T cell production leads to an increased incidence of viral diseases and certain types of cancer.

The most significant example of an acquired immunodeficiency is **acquired immunodeficiency syndrome** (**AIDS**). AIDS is not a single disease, but a syndrome—a group of signs, symptoms, and diseases associated with a common pathology. AIDS is characterized by the presence of several rare or opportunistic infections along with infection by human immunodeficiency virus (HIV) or as a severe decrease in CD4 cells and a test showing the presence of HIV.

Human immunodeficiency virus (**HIV**) is an enveloped, +ssRNA retrovirus that uses reverse transcriptase to synthesize a dsDNA copy of its genome. Its envelope is characterized by a large glycoprotein named **gp120**, which is its primary attachment molecule, and by a smaller glycoprotein named **gp41** that promotes fusion of the viral envelope to a target cell.

HIV replicates in seven steps:

1. *Attachment.* HIV primarily attaches to four cell types: helper T cells, cells of the macrophage lineage, smooth muscle cells, and dentritic cells.

2. *Entry and uncoating*. Following fusion of the viral envelope with the cell's cytoplasmic membrane, the intact viral capsid enters the cell and the virus uncoats to release its ssRNA into the cell's cytoplasm.
3. *Synthesis of DNA*. Reverse transcriptase transcribes a dsDNA copy of its +ssRNA g enome.
4. *Integration*. The dsDNA provirus inserts into a human chromosome and remains a part of the cellular DNA for life.
5. *Synthesis of RNA and polypeptides*. An infected cell transcribes and translates the integrated provirus and makes viral genomes and polypeptides.
6. *Release*. Molecules of genomic RNA and tRNA bud from the cytoplasmic membrane.
7. *Assembly and maturation*. After the virion buds from the cells, the viral enzyme **protease** promotes final assembly of a virulent HIV.

Progressive death of helper T (CD4) cells allows opportunistic pathogens to proliferate, resulting in the characteristic signs and symptoms of AIDS. **Antiretroviral therapy (ART)**, a "cocktail" of three to four antiviral drugs, is the standard treatment for AIDS. Progress in developing a vaccine against HIV has been disappointing. Because HIV is typically spread through sexual contact and sharing needles, preventive measures include abstinence, condom usage, use of clean, sterile needles, and screening of blood and donated organs.

KEY THEMES

The immune system is our ultimate protection against pathogens. It is a specialized system geared toward destroying foreign invaders. When properly functioning, it is more effective at protecting us than all of our best technology. When immune control lapses, however, the results can be devastating. As you read, keep in mind that malfunctions can occur in one of two ways:

Hypersensitivities and autoimmune diseases result in too much or uncontrolled cell death: Both of these types of reactions are overreactions—responses to antigens that aren't harmful or responses that haven't been turned off. Both lead to more damage than is warranted.

Immunodeficiencies result in too little cell death: Depressed immune responses occur for a variety of reasons, but the end result is that the body loses its best defense in the face of invading pathogens. Without the strength of the immune system to counter the activity of microbes, the microbes multiply unchecked, entirely to our loss.

QUESTIONS FOR FURTHER REVIEW

Answers to these questions can be found in the answer section at the back of this study guide. Refer to the answers only after you have attempted to solve the questions on your own.

Multiple Choice

1. Allergic responses are examples of what type of hypersensitivity reaction?
   a. type I
   b. type II
   c. type III
   d. type IV

2. In allergic individuals, exposure to allergens stimulates B cells to produce ____________________.
   a. IgE
   b. IgG
   c. IgM
   d. IgA
3. The primary chemical released by mast cells during degranulation is of what class?
   a. kinins
   b. leukotrienes
   c. IgE
   d. histamine
4. Type II hypersensitivities involve
   a. improper complement and antibody responses.
   b. autoimmune disorders.
   c. improper T cell responses.
   d. both a and b.
5. Which antigen is most responsible for provoking type II hypersensitivities?
   a. AB antigens
   b. Rh antigens
   c. allergens
   d. All of the above cause type II hypersensitivities.
6. Graft-versus-host disease is due to which type of hypersensitivity reaction?
   a. type I
   b. type II
   c. type III
   d. type IV
7. Which of the hypersensitivity reactions is most widely found in the general population?
   a. type I
   b. type II
   c. type III
   d. type IV
8. Which of the following immunosuppressive drugs would be prescribed to prevent allograft rejection?
   a. cyclophosphamide
   b. cyclosporine
   c. corticosteroids
   d. Both b and c together should be prescribed.
9. Which of the following immunosuppressive drugs would be most beneficial for someone with asthma?
   a. cyclosporine
   b. corticosteroids
   c. antihistamines
   d. None of the above should be prescribed.
10. Autoimmune diseases
    a. can arise as a result of environmental factors.
    b. can arise as a result of genetic factors dealing with molecules such as MHC.
    c. can arise as a result of molecular mimicry in infectious agents such as bacteria.
    d. All of the above can lead to autoimmune diseases.
11. An individual who has an autoimmune disease that affects his or her blood cells most likely would NOT experience problems with
    a. combating infections.
    b. transporting oxygen.
    c. clotting.
    d. The individual would have difficulty with all of these functions.

12. The autoimmune disease lupus is an example of what type of hypersensitivity?
    a. type I
    b. type II
    c. type III
    d. type IV
13. Which of the diseases below is NOT a primary immunodeficiency?
    a. SCID
    b. IgA deficiency
    c. HIV
    d. DiGeorge syndrome
14. Immunodeficiencies are
    a. always a result of genetic problems in the host.
    b. always a result of genetic properties of the invading microorganism.
    c. always a result of environmental factors.
    d. a result of all three of the above, depending on the actual disease.
15. The most common mode of transmission for HIV is
    a. sexual contact.
    b. intravenous drug abuse.
    c. accidental needlesticks.
    d. mother to child by breast milk.

Fill in the Blanks

1. If the immune system overreacts to perceived antigens, this is a(n) ____________________ response. If the immune system underreacts, this is an example of ____________________.
2. Antibodies produced during allergic reactions sensitize three types of defense cells to respond to allergens: ____________________, ____________________, and ____________________.
3. Type II hypersensitivity reactions against small drug molecules bound to red blood cells are called ____________________.
4. The four types of grafts, as distinguished by relatedness to the host, are ____________________, ____________________, ____________________, and ____________________. Of these four, ____________________ are the most likely to illicit an immune response so severe as to cause loss of the graft.
5. Autoimmune diseases predominantly involve ____________________ (B/T) cell-mediated response irregularities. Such diseases can be of two types: ____________________ or ____________________ autoimmune diseases.

6. The most important autoimmune disease of nervous tissue is

   ____________________.

7. ____________________ are the hallmarks of immunodeficiency.

8. Primary immunodeficiences are usually due to ____________________ or

   ____________________ defects; acquired immunodeficiences are usually

   the result of other parameters such as ____________________,

   ____________________, and ____________________.

9. ____________________ is the type of HIV that is more prevalent in the

   United States.

Short-Answer Questions for Thought and Review

1. What triggers degranulation of sensitized cells during allergic reactions? How does degranulation account for the symptoms typical of "hay fever"?

2. What is the essential difference between the small molecule antigen response elicited in type II hypersensitivity reactions and that elicited in type IV hypersensitivity reactions?

3. Why would autoimmune diseases occur more frequently in older individuals?

Critical Thinking

1. Your roommate goes out for a walk in the woods and comes back complaining about being bitten by fire ants. Her leg is already swelling around the bites, which are red and itchy. She applies a topical agent containing an antihistamine and an analgesic for the itching, but it does little to stop or reverse the inflammation. Additionally, the analgesic wears off quickly. How do you explain to her why her topical cream didn't work?
2. Individuals with type AB blood are designated universal recipients; those with type O blood are universal donors. For which group of individuals could prior exposure to certain bacterial antigens result in incompatible blood donations? Why? What is the ultimate danger of such infections in the human body?
3. Immunosuppressive drugs are vital to the success of tissue grafts. What, however, is the major disadvantage of using these drugs long term? Why?

4. Can an autoimmune disease also be an immunodeficiency disease? Answer yes or no, and explain.

Concept Building Questions

1. Based on the material we have studied so far in this and previous chapters, explain, using the concepts of genetics and mechanisms of adaptive and innate immunity, why immunotherapy for allergies works only for two or three years before it has to be done again. Why aren't these treatments permanent?
2. Explain why privileged sites are a benefit for graft transplantation but a significant detriment if the same sites become colonized by pathogenic microbes. Use the material you have learned in other chapters on immunity and in the chapters on the organisms themselves to help you answer this question. Why aren't there more infections in privileged sites?
3. Based on what you have learned so far in this text about viruses, why might viruses be responsible for many immune function disorders such as autoimmunity?

19 Microbial Diseases of the Skin and Wounds

CHAPTER SUMMARY

Structure of the Skin (pp. 556–557)

The skin is composed of two distinct layers: The deeper **dermis** is a tough layer of loosely packed cells, protein and muscle fibers, glands, blood vessels, nerve endings, and hair follicles. It provides strength and flexibility. The superficial **epidermis,** which is continually replaced, is a thin layer of tightly packed cells filled with a waterproofing protein called *keratin.*

Several factors make the surface of the skin inhospitable to microbes. Defensive *dendritic cells* phagocytize microbes and deliver microbial antigens to defensive lymphocytes. Sweat contains salt and other antimicrobial chemicals. *Sebum,* an oily lipid secreted by sebaceous glands in the dermis, also has antimicrobial properties.

Wounds are trauma to any tissue of the body. Wounds can breach the mechanical barrier provided by intact epidermis and dermis, allowing microbes to infect the deeper tissues of the body. Dirty wounds provide platforms for the growth of biofilms. Some wound infections overwhelm the body's defenses and result in severe diseases.

Normal Microbiota of the Skin (p. 557)

Microbiota are normally harmless residents of the body. Significant skin microbiota include small lipophilic yeasts such as *Malassezia*, which digest sebum, as well as bacteria in the genera *Staphylococcus* and *Micrococcus*. **Diphtheroids** such as *Propionibacterium acnes* are also common Gram-positive bacterial microbiota of the skin.

Bacterial Diseases of the Skin and Wounds (pp. 557–570)

Common bacteria infecting the skin are *Staphylococcus, Streptococcus,* and *Propionibacterium.*

Folliculitis

Folliculitis is an infection of a hair follicle that causes the base of the follicle to become red, swollen, and pus filled. On the skin, this infection is commonly called a **pimple,** whereas at the base of an eyelid it is called a **sty.** A **furuncle** (*boil*) results when the infection spreads to surrounding tissue. When several furuncles join, a **carbuncle** is formed.

Staphylococcus epidermis and *Staphylococcus aureus* are the most common causes of folliculitis and associated infections of the skin. Both produce polysaccharide slime layers that inhibit phagocytosis and toxins that contribute to their pathogenicity. Transmission is via direct contact and fomites. Treatment is with a semisynthetic form of penicillin called didoxacillin; however, resistant *Staphylococcus aureus* has emerged as a major health problem. Vancomycin is used to treat such infections, but vancomycin resistance is rising. Aseptic techniques are especially important in hospitals to protect patients from methicillin-resistant *S. aureus* (MRSA).

Staphylococcal Scalded Skin Syndrome

In **staphylococcal scalded skin syndrome,** infection with a strain of *S. aureus* causes cells of the epidermis to separate from one another and from the underlying tissue. Within two days, the epidermis peels off in sheets, as if scalded. The reaction is due to **exfoliative toxins** that are secreted by about 5% of strains of *S. aureus*. Transmission is from person to person via contact with cut or abraded skin. Treatment involves administration of methicillin or another antimicrobial.

Impetigo (Pyoderma) and Erysipelas

Impetigo, also called **pyoderma,** is a contagious skin disease caused by *S. aureus* alone (about 80% of cases), or by *Streptococcus pyogenes* alone or in conjunction with *S. aureus*. It is characterized by the presence of small, red patches that develop into oozing, pus-filled vesicles. These vesicles break to form a honey-colored crust that is attached firmly to the skin. Preschool children, whose immune systems are immature, are most commonly affected. **Erysipelas** results when infection spreads to lymph nodes, causing pain, fever, chills, and leukocytosis. Left untreated, erysipelas may be fatal. Some strains of *S. pyogenes* may spread via the blood to the kidneys, causing glomerulonephritis. Treatment involves administration of antibiotics and regular, gentle cleansing of affected areas.

Necrotizing Fasciitis

S. pyogenes can cause necrotizing soft tissue infection, commonly called "flesh-eating strep" and clinically known as necrotizing fasciitis. The disease is characterized by intense sensitivity, pain, and swelling at the infection site, followed by bacterial digestion of the muscle fascia and adipose tissue, which causes the overlying skin to become distended and discolored. Flulike symptoms, mental confusion, and toxemia develop. Organ failure and death are common.

Strains of *S. pyogenes* that cause necrotizing fasciitis produce enzymes that assist the bacterium to invade body tissues, and secrete potent toxins that destroy tissues. The usual route of transmission is from person to person, most commonly following a cut or other injury. Treatment involves aggressive removal of affected tissue, plus a combination of clindamycin and penicillin.

Acne

Acne is most commonly caused by propionibacteria, which are small, Gram-positive rod-shaped diphtheroids normally found on the skin. Excessive oil production, typically triggered by the hormones of adolescence, stimulates their growth. The bacteria secrete chemicals that attract leukocytes and trigger inflammation. Acne is thus characterized by the presence of pimples filled with dead bacteria and living leukocytes. When dead bacteria block pores, blackheads result. Acne is treated with topical antimicrobial agents, and with the Clear Light system, which uses a blue light wavelength to destroy the bacteria. Severe cases of acne are treated with Accutane, a derivative of vitamin A; however, the drug can cause intestinal bleeding and birth defects and is not prescribed to women who are or may become pregnant.

Cat Scratch Disease

Bartonella henselae, a Gram-negative aerobic bacillus, causes **cat scratch disease,** a condition characterized by a few days of fever and malaise, accompanied by localized swelling at the site of infection and in nearby lymph nodes for several months. The bite or scratch of an infected cat or a flea introduces the bacterium into the skin. The disease is caused by endotoxin that is released when the bacterial cells die. Treatment is with antimicrobial drugs.

Pseudomonas Infection

Pseudomonas aeruginosa is a Gram-negative, aerobic bacillus that is almost everywhere in soil, decaying organic matter, and many hospital fomites, including sinks, damp sponges, dialysis machines, and respirators. Harmless to patients with intact skin, it typically gains access to the fascia and deeper tissues of burn victims to produce invasive infection and shock. It can also invade the external ear canal of swimmers to cause **otitis externa.** The bacterium produces a blue-green pigment called **pyocyanin,** as well as toxins and enzymes that contribute to host tissue damage. Because the bacterium is resistant to a wide variety of antimicrobials, physicians treat infections with a combination of drugs. Polymixin, which is toxic to humans, is used as a last resort.

Rocky Mountain Spotted Fever

Infection with *Rickettsia rickettsii* causes **Rocky Mountain spotted fever (RMSF),** which is characterized by a rash, flulike symptoms, system failures, encephalitis, subcutaneous hemorrhages called **petechiae,** and death in about 5% of patients. The bacterium is carried by ticks, and is released into the host's circulatory system following a bite. Early diagnosis and treatment with doxycycline, tetracycline, or chloramphenicol is crucial. No vaccine is available.

Cutaneous Anthrax

Anthrax has three distinct clinical manifestations: gastrointestinal, inhalation, and cutaneous. All are caused by *Bacillus anthracis.* **Cutaneous anthrax** results when the bacterium enters a wound in the skin, producing a solid skin nodule that spreads to form a painless, swollen, black ulcer called an **eschar.** Treatment with antimicrobial drugs is usually successful. A safe and effective vaccine is available, but requires six doses over 18 months plus annual boosters.

Gas Gangrene

Ischemia develops when blood supply to a tissue is interrupted. The tissue becomes anaerobic and necrosis (death) occurs. **Gas gangrene** can develop in the dead tissue if it is infected with endospores of anaerobic *Clostridium.*

Viral Diseases of the Skin and Wounds (pp. 571–582)

Viral diseases often manifest signs and symptoms affecting the skin.

Diseases of Poxviruses

Poxvirus causes smallpox, the first human disease to be eradicated globally in nature. It also causes three diseases of animals: orf, cowpox, and monkeypox. These can be transmitted to humans who have prolonged, close contact with infected animals. All poxviruses produce lesions that progress through a series of stages: from flat **macules** to raised **papules** to fluid-filled **vesicles** to pus-filled **pustules** (also known as **pox** or pocks). The pustules dry to form crusts.

Smallpox virus, commonly called **variola,** exists in two strains: variola major and variola minor. Historically, variola major produced death in about 20% to 40% of patients. Stocks of variola are still maintained in laboratories and could be used as a biological weapon; thus, US military personnel and some health care providers are now routinely immunized.

Herpes Infections

Herpes simplex viruses type 1 (HHV-1) and type 2 (HHV-2) cause the painful, itchy skin lesions of **herpes** (commonly called fever blisters or cold sores). Transmission is most commonly via skin contact; however, asymptomatic carriers can shed HHV-2 genitally. After entering the body through broken mucous membranes, the virus reproduces in epithelial cells. **Herpetic gingivostomatitis** is a severe infection in the oral cavity. Young adults with sore throats from other viral infections may develop **herpetic pharyngitis.** If herpes enters a cut in the finger, a **whitlow** (herpetic blister) may develop. **Herpes gladiatorum** can develop following skin contact with herpes lesions during wrestling.

After a primary infection, herpes simplex viruses often enter sensory nerves, where they remain latent and can become reactivated. Fortunately, recurrent lesions are rarely as severe as those of initial infections. Treatment is with antiviral medications such as acyclovir. Because herpes is life threatening in newborns, the babies of women with genital lesions are delivered via cesarean section.

Warts

Warts (or **papillomas**) are benign growths of the epithelium of the skin or mucous membranes. Genital warts may grow to become giant protrusions called **condylomata acuminata.** Almost 60 different strains of *Papillomavirus* cause warts. The virus is transmitted via direct contact and fomites. Most warts, although painful and unsightly, are harmless. However, some genital warts are associated with an increased risk of cancer, including penile, vaginal, and cervical cancer.

Diagnosis of warts is usually via simple observation, and treatment may include the use of caustic chemicals, surgery, freezing, cauterization, or laser treatment. Prevention of most warts is difficult, though genital warts can be prevented by abstinence and mutual monogamy.

Chickenpox and Shingles

Varicella-zoster virus (VZV) causes **chickenpox** (clinically called **varicella**). About two to three weeks after infection with the virus, patients, typically children, develop a fever and characteristic itchy rash. In about 15% of individuals who have had chickenpox, stress, aging, or immune suppression causes the viruses to reactivate, travel down the nerve they inhabit, and produce the extremely painful band of lesions known as **shingles** (also called **herpes zoster**).

Uncomplicated chickenpox is self-limiting and requires no treatment. Acyclovir provides relief from the painful rash of shingles for some patients. The

CDC recommends immunization against chickenpox for all children between 12 and 18 months of age and a second dose before starting school.

Rubella

Rubella (also called German measles and three-day measles) is a generally harmless disease in children, causing only slightly swollen lymph nodes and a mild rash that lasts about three days. Infections of pregnant women, however, result in a range of birth defects, including cardiac abnormalities, deafness, blindness, mental retardation, and others. Death of the fetus is also common. The infection is caused by *Rubivirus,* which spreads via respiratory secretions and infects cells of the upper respiratory tract to cause errors of DNA replication in dividing cells. No treatment is available. Rubella vaccine is given during early childhood, but is made from a live, attenuated virus, and therefore should never be given to pregnant women or immunocompromised patients.

Measles (Rubeola)

Measles, also known as **rubeola,** is one of the more contagious and serious of the childhood viral diseases. Measles begins with fever and malaise, followed 2 days later by Koplik's spots on the mucous membranes of the mouth. These fade within one to two days and are followed by red, raised body lesions. Complications include pneumonia, encephalitis, and **subacute sclerosing panencephalitis (SSPE),** a progressive disease of the central nervous system.

Morbillivirus, which causes measles, is a single-stranded RNA virus with a protein that triggers fusion between an infected cell and its neighbors, allowing the virus to travel undetected by the host's immune system. Most patients recover within two to three weeks, but measles can cause death in the immunocompromised. No treatment is available. Vaccination has nearly eliminated measles as an endemic disease in the United States, but it is a frequent cause of death in other countries.

Other Viral Rashes

Erythema infectiosum (also called **fifth disease**) is a respiratory disease caused by a ssDNA virus in the genus *Erythrovirus.* It causes a mild rash that begins on the face and spreads to the trunk and limbs. No treatment is available. **Roseola** is an endemic illness of children caused by a herpesvirus in the genus *Roseolovirus.* It causes fever, sore throat, enlarged lymph nodes, and a faint pink rash. Several types of **coxsackie A viruses** cause self-limiting lesions of the mouth and pharynx called **herpangina** because they resemble herpes lesions. Another serotype causes hand-foot-and-mouth disease, which involves lesions on the extremities and in the mouth.

Mycoses of the Hair, Nails, and Skin (pp. 582–587)

Mycoses are fungal diseases. The fungi that cause mycoses are mainly opportunistic.

Superficial Mycoses

Superficial mycoses are confined to the hair, nails, and outer layers of the skin, all of which are composed of dead cells filled with keratin. Superficial fungi produce **keratinase,** an enzyme that dissolves keratin. Examples are black and white **piedra,** which is caused by ascomycetes that invade hair shafts, and **pityriasis versicolor,** also called *versicolor,* which is caused by a basidiomycete called *Malassezia furfur.* In pityriasis, fungal interference with melanin production causes hypo- or hyperpigmented patches of skin.

Cutaneous Mycoses

Dermatophytoses are cutaneous infections caused by dermatophytes, fungi that feed on keratin and colonize dead layers of skin, nails, and hair. Colonization stimulates cell-mediated immune responses that damage deeper tissues. Common dermatophytoses are tinea pedis (athlete's foot), tinea cruris (jock itch), and tinea unguium (onychomycosis). Limited infections can be treated topically, but widespread and nail infections must be treated with oral antifungal agents.

Wound Mycoses

Some fungi grow on deeper tissues in the body and eventually grow up into the epidermis to produce lesions on the skin's surface.

Four species of ascomycete fungi cause **chromoblastomycosis,** which initially manifests as scaly lesions on the skin but progresses to large, wartlike lesions that grow to extensive tumors. Treatment involves removal of infected and surrounding tissue and oral antifungal agents.

Phaeohyphomycosis is acquired when fungal spores, which are prevalent within buildings such as hospitals, enter open wounds. Some cases can be treated with itraconazole, but the disease is permanently destructive to affected tissues.

Mycetomas are tumorlike infections of the skin, fascia, and bones of the hands or feet caused by soil fungi of several genera in the division Ascomycota. Treatment combines surgery and antifungal therapy with ketoconazole for one to three years.

Sporotrichosis is a subcutaneous infection usually limited to the arms and legs that occurs when *Sporothrix schenckii* is introduced into wounds, often by thorn pricks or wood splinters. Cutaneous lesions can be treated successfully with topical applications of saturated potassium iodide for several months.

Parasitic Infestations of the Skin (pp. 587–589)

Some parasitic protozoa and arthropods infect the skin and cause diseases.

Leishmaniasis

Leishmania is a parasitic protozoan transmitted to humans by sand flies. It causes any of three forms of **leishmaniasis:** cutaneous leishmaniasis leaves permanent scars; mucocutaneous leishmaniasis causes permanently disfiguring damage to the membranes of the mouth, nose, or soft palate; visceral leishmaniasis occurs when macrophages spread the infection to body organs, and is fatal in 95% of untreated cases. Treatment for severe infections requires administration of sodium stibogluconate, an antimicrobial drug not licensed for use in the United States.

Scabies

The mite *Sarcoptes scabiei* causes **scabies,** a disease characterized by intensely itchy lesions, by burrowing into human skin and triggering inflammation. The mite is spread via close contact or shared fomites. Scratching can introduce bacteria into the wounds, triggering impetigo. Treatment involves the application of a miticide lotion, as well as washing and drying all fomites.

QUESTIONS FOR FURTHER REVIEW

Answers to these questions can be found in the answer section at the back of this study guide. Refer to the answers only after you have attempted to solve the questions on your own.

Multiple Choice

1. Which of the following is not a component of the dermis?
   a. sweat glands
   b. blood vessels
   c. basal cells
   d. hair follicles

2. _____ and _____ are diseases of the skin caused by bacteria.
   a. Impetigo, warts
   b. Measles, cutaneous anthrax
   c. Measles, scabies
   d. Acne, cutaneous anthrax

3. Which of the following is true about *Staphylococcus* species?
   a. gram negative
   b. aerobic
   c. rod shaped
   d. salt tolerant

4. Which of the following bacteria is an example of a diphtheroid?
   a. *Staphylococcus aureus*
   b. *Propionibacterium acnes*
   c. *Bacillus anthracis*
   d. *Rickettsia rickettsii*

5. *Pseudomonas aeruginosa* is likely to be found in the following places.
   a. soil
   b. swimming pools
   c. hot tubs
   d. All the above are correct.

6. The vector for *Rickettsia rickettsii* is the
   a. tick.
   b. flea.
   c. spider.
   d. beetle.

7. Which of the following is not a DNA virus?
   a. rubella virus
   b. human herpes virus 1
   c. varicella-zoster virus
   d. variola virus

8. Human herpes viruses 1 and 2 cause infections at which of the following sites?
   a. lips
   b. eyes
   c. finger
   d. All the above are correct.

9. Which of the following treatments should not be administered to a young child with chickenpox?
   a. antihistamines
   b. acetaminophen
   c. aspirin
   d. None of the above is correct.

10. For which of the following viral diseases is a vaccine not available?
    a. chickenpox
    b. rubella
    c. smallpox
    d. human herpes virus 2

11. Anthropophilic dermatophytes are associated with
    a. humans only.
    b. animals only.
    c. both animals and humans.
    d. neither animals or humans.

12. The form of leishmaniasis that is most often fatal is
    a. cutaneous leishmaniasis.
    b. mucocutaneous leishmaniasis.
    c. visceral leishmaniasis.
    d. none of the above.

13. Ringworm is a term formerly used to describe
    a. phaeohyphomycoses.
    b. dermatophytoses.
    c. mycetomas.
    d. chromoblastomyces.
14. Human herpes viruses 1 and 2 remain latent in what cell type?
    a. epithelial cells
    b. endothelial cells
    c. nerve cells
    d. hepatic cells
15. Which of the following is a problem of viral latency?
    a. Latent viruses can reactivate to make the host sick years after initial exposure.
    b. Latent viruses can integrate into the host genome and cause cancers.
    c. Latent viruses cannot be cleared permanently from the host and remain with the host for life.
    d. All of the above are problems of viral latency.
16. Vaccination against which of the following diseases is not included in the MMR vaccine?
    a. measles
    b. roseola
    c. rubella
    d. mumps
17. Mycoses are diseases caused by
    a. viruses.
    b. bacteria.
    c. plants.
    d. fungi.
18. Pityriasis is a fungal infection that interferes with production of
    a. melanin.
    b. sebum.
    c. basal cells.
    d. keratin.

Fill in the Blanks

1. Bacteria that are normal residents of the skin and usually harmless are considered ____________________.
2. The darkness of one's skin is determined by the amount of ____________________ present.
3. The outermost layer of the skin is referred to as the ____________________.
4. ____________________ is an enzyme produced by *S. aureus* that causes blood to clot.
5. *S. aureus* that are resistant to methicillin are treated with the antibiotic ____________________.
6. The causative agents of impetigo are ____________________ and ____________________.

7. Mycoses are classified as ____________________,

____________________, ____________________,

or____________________.

8. The virulence factor lipid A is also commonly referred to as

____________________.

9. Genital herpes is primarily caused by ____________________.

10. Varicella-zoster virus is the causative agent of both ____________________

and ____________________.

11. Varicella-zoster virus is transmitted primarily by ____________________

and ____________________.

12. Fifth disease is another name for ____________________.

13. Coxsackie A virus can cause herpeslike lesions called

____________________.

14. Subacute sclerosing panencephalitis is a complication associated with

____________________.

Matching

Match the disease on the left to the causative agent on the right. Each letter will be used only once.

1. ____Mycetomas
2. ____Erythema infectiosum
3. ____Scalded skin syndrome
4. ____Roseola
5. ____Necrotizing fasciitis
6. ____Piedra
7. ____Cutaneous anthrax
8. ____Scabies
9. ____Chickenpox
10. ____Athlete's foot

A. *Trichophyton rubrum*
B. Exophiala
C. *Sarcoptes scabiei*
D. *Staphylococcus aureus*
E. *Trichosporon beigelii*
F. *Streptococcus pyogenes*
G. Human herpesvirus 6
H. *Bacillus anthracis*
I. Erythrovirus
J. Varicella-zoster virus

Short-Answer Questions for Thought and Review

1. Explain why the skin surface is typically an inhospitable environment for many microbes.

2. How can the presence of a slime layer (capsule) be a virulence factor for the bacterium that has one?

3. Why does *Pseudomonas aeruginosa* rarely cause disease despite its widespread occurrence and various virulence factors?

4. What are the virulence factors that contribute to the ability of certain *S. pyogenes* strains to cause necrotizing fasciitis?

Critical Thinking

1. Humans can be infected with various poxviruses such as variola virus as well as cowpox and monkeypox. However, infections due to cowpox or monkeypox are far less common than those once caused by variola virus. Explain.
2. An individual has acquired genital herpes from exposure to human herpes virus. He is placed on chemotherapeutic agents as treatment and the symptoms subside. However, over time the lesions return. Must these new lesions have been caused from new exposure to the human herpes virus?
3. A couple's child has acquired chickenpox from a playmate at school. Is it possible for the parents to develop shingles if exposed to the virus responsible for their child's illness? Explain.

Concept Building Questions

1. Why is it often necessary to treat a dermatophytotic nail infection with oral medications rather than topical agents?
2. There is quite a bit of debate over whether or not the last smallpox stocks should be destroyed. Give one reason why they should be destroyed, and one reason why they should not be destroyed.

20 Microbial Diseases of the Nervous System and Eyes

CHAPTER SUMMARY

Structure of the Nervous System (pp. 598–600)

The nervous system is divided into the central nervous system and the peripheral nervous system.

Structures of the Central Nervous System

The **central nervous system (CNS)** is composed of the brain and spinal cord surrounded by three layers of tissues called **meninges**: an outermost **dura mater**, a middle **arachnoid mater**, and an inner tissue closely appressed to the spinal cord and brain, called the **pia mater**. Blood vessels that supply the CNS are composed of tightly joined cells that form the **blood-brain barrier**, which prevents most microbes and large molecules in the blood from entering the **cerebrospinal fluid (CSF)** that fills the subarachnoid space. CSF acts as a shock absorber, provides nutrients and oxygen to the CNS, and removes wastes. CSF can be removed from the lumbar region of the spinal cord in a medical procedure called a **spinal tap**.

Structures of the Peripheral Nervous System

The **peripheral nervous system (PNS)** is composed of **motor nerves** that transfer commands from the CNS to muscles and other organs throughout the body, and **sensory nerves** that transmit information to the CNS concerning events in the body and environmental stimuli. Mixed nerves carry signals both toward and away from the CNS. Branches of nerves often merge together to form a nerve **plexus**.

Cells of the Nervous System

The entire nervous system is composed of two types of cells: **Neuroglia** are supportive cells that provide scaffolding, insulation, nutrition, or defense. **Neurons** are cells that generate an electrical signal called an *action potential* or *nerve impulse*. The cell bodies of neurons contain the nucleus. Grouped together, nerve cell bodies form a **ganglion**. Numerous short **dendrites** and a longer single **axon** extend from a cell body; the dendrites carry impulses toward the cell body and the axon carries impulses away. Within an axon, the cytoskeleton transports substances by a process known as **axonal transport**. The terminal ends of axons have branches that form junctions called **synapses** with gland cells, muscle cells, and other neurons. The space between an axon and a neighboring cell, called a **synaptic cleft**, stops the transmission of electrical signals. To bridge the gap, an axon terminal releases chemical messengers called **neurotransmitters** into the synaptic cleft. The binding of these molecules to receptors on target cells initiates impulses.

Portals of Infection of the Central Nervous System

The central nervous system is an *axenic* environment: it has no normal microbiota. Microbes carried in the blood or lymph may penetrate the blood-brain barrier by infecting and killing cells of the meninges, causing **meningitis.**

Bacterial Diseases of the Nervous System (pp. 600–611)

Bacteria and their toxins can cause disease of the nervous system.

Bacterial Meningitis

Bacterial meningitis is inflammatory bacterial infection of the meninges, most commonly the pia and arachnoid mater. It is characterized by an increased number of white blood cells in the CSF, sudden high fever, and severe headache, stiff neck, nausea, vomiting, pain, and in many cases drowsiness, confusion, and irritability, and sometimes **petechiae**, small hemorrhages of blood vessels in the skin. Infection of the brain, called *encephalitis*, can result in death.

More than 50 species of bacteria can cause meningitis; however, most cases are caused by *Neisseria meningitidis, Streptococcus pneumoniae*, *Haemophilus influenzae*, *Listeria monocytogenes*, and *Streptococcus agalactiae*, the latter most commonly in newborns. In most cases, bacteria spread to the meninges from infections of the lungs, sinuses, or inner ear via the blood (bacteremia). However, *Listeria* enters the body in contaminated food or drink. It causes a form of meningitis called **listeriosis**. For all forms, prompt antimicrobial therapy is vital.

Hansen's Disease (Leprosy)

Hansen's disease (**leprosy**) is caused by infection with *Mycobacterium leprae*. Patients with a strong T cell immune response develop a nonprogressive form of the disease called tuberculoid leprosy, which is characterized by limited nerve damage and loss of sensation. In contrast, patients with a weak T cell immune response develop lepromatous leprosy, in which the bacterium multiplies in skin, mucous membranes, and nerve cells, progressively destroying facial features, fingers, toes, and other body structures. The cell wall of *Mycobacterium leprae* contains a large amount of mycolic acid, which accounts for the slow growth rate of the bacterium and protects it from lysis once phagocytized. Diagnosis of leprosy is based on loss of sensation, disfigurement, and the presence of **acid-fast bacilli** (**AFB**) in samples from affected sites. Therapy requires administration of multiple drugs for a minimum of two years. The tuberculosis vaccine (BCG) provides some protection.

Botulism

Botulism is not an infection, but instead an intoxication (poisoning) caused by a **neurotoxin** of *Clostridium botulinum* that progressively paralyzes skeletal muscles. **Food-borne botulism**, **infant botulism**, and **wound botulism** are three forms, all of which are characterized by the presence of botulism neurotoxins that prevent muscle contraction by inhibiting the secretion of the neurotransmitter **acetylcholine** in the **neuromuscular junction** (the synapse between a motor neuron and a muscle cell). Treatment of botulism entails: (1) repeated washing of the intestinal tract to remove *Clostridium*, (2) administration of botulism immune globulin intravenously (BIG-IV), and (3) administration of antimicrobial drugs, in infant and wound botulism.

Tetanus

Another *Clostridium* species, *C. tetani*, causes **tetanus.** The classic symptom of tetanus is severe muscle contraction, initially affecting the jaw and neck muscles (called *lockjaw*). This is caused by a potent neurotoxin called **tetanospasmin** which blocks inhibitory neurotransmitters, resulting in unrelenting contraction. Other symptoms include sweating, drooling, and irritability. If untreated, patients die when constant contraction of the diaphragm muscle makes breathing impossible.

C. tetani typically enters the body through a break in the skin or mucous membranes and travels over days or weeks to the central nervous system. Treatment involves thorough cleansing of all wounds and immediate passive immunization in which immunoglobulin directed against tetanospasmin is injected, as well as administration of antimicrobials. The CDC recommends immunization with tetanus toxoid beginning at 6 months of age followed by a booster every 10 years for life.

Viral Diseases of the Nervous System (pp. 611–618)

Because viruses are smaller than cells, they can more readily cross the blood-brain barrier; therefore, there are more viral infections of the nervous system than bacterial or fungal infections.

Viral Meningitis

Viral meningitis (*aseptic meningitis*) is more common than bacterial or fungal meningitis, and is usually milder. Although it also causes severe headache, stiff neck, and other symptoms associated with bacterial meningitis, death from viral meningitis is rare. About 90% of cases result from infections of **enteroviruses,** which are transmitted via fecal contamination of food, water, or hands and spread via the bloodstream to infect the meninges. No specific treatment exists.

Poliomyelitis

Infection with poliovirus, another *Enterovirus*, causes any one of four forms of **poliomyelitis** (or **polio**): (1) asymptomatic infection (about 90%); (2) minor polio, which is characterized by temporary fever, headache, malaise, and sore throat; (3) nonparalytic polio, which results from infection of the meninges and produces symptoms of minor polio as well as muscle spasms and back pain; and (4) paralytic polio, which results from invasion of the spinal cord and cerebrum and produces varying degrees of paralysis. **Postpolio syndrome** is a muscle deterioration that affects polio patients 30–40 years after their original illness. People typically get poliovirus by drinking contaminated water. There is no specific treatment. Two effective vaccines make worldwide polio eradication possible.

Rabies

Rabies virus is a negative, ssRNA virus in the genus *Lyssavirus*. It causes **rabies,** a degenerative neurologic disease characterized by pain or itching at the site of infection, fever, headache, malaise, anorexia, hydrophobia (triggered by painful swallowing), seizures, disorientation, paralysis, and death. Rabies is a **zoonosis;** that is, a disease spread from animal reservoirs to humans. Bats, dogs, cats, and other animals may carry the virus in their saliva. Initial treatment of any suspect wound is critical, and includes thorough washing and application of antirabies serum.

Injection of human rabies immunoglobulin (HRIG) is followed by a series of injections with rabies vaccine that allow the patient to build effective immunity before the disease develops. If treatment is delayed until symptoms appear, the disease is almost invariably fatal.

Arboviral Encephalitis

*Ar*thropod-*bo*rne *viruses,* called **arboviruses**, are transmitted between hosts by blood-sucking arthropods. Mosquitoes transmit various types of **arboviral encephalitis**, which affects birds, horses, rodents, and rarely humans. Examples are Eastern equine encephalitis and West Nile virus. Typically, infections result in only mild, cold-like symptoms; occasionally, arboviruses in the blood cross the blood-brain barrier and cause encephalitis, from which the neurological damage may be permanent. Treatment is supportive. No human vaccine is currently available.

Mycoses of the Nervous System (pp. 618–620)

Fungi rarely infect the central nervous system. Occasionally, **mycoses** (fungal diseases) spread from the lungs via the blood to the meninges.

Cryptococcal Meningitis

Cryptococcus neoformans is a basidiomycete yeast that causes **cryptococcal meningitis**. The disease manifests with the signs and symptoms common to other forms of meningitis. Human infections begin with inhalation of spores and/or dried yeast cells from airborne bird droppings. In some patients, the fungus spreads from the lungs via the blood to the meninges and brain tissue. Infections often appear in AIDS patients and transplant recipients taking immunosuppressive drugs. Treatment is with a combination of antimicrobial drugs. No vaccine is available.

Protozoan Diseases of the Nervous System (pp. 620–622)

Protozoan infections of the nervous system are relatively rare.

African Sleeping Sickness

The protozoan *Trypanosoma brucei* causes **African sleeping sickness**. This parasite randomly changes its surface glycoprotein antigens each time it replicates, leaving the host's immune system unable to clear the infection and develop immunity. African sleeping sickness is characterized by fever, swollen lymph nodes, and headache, followed by meningoencephalitis and extreme drowsiness, coma, and death. The blood-sucking tsetse fly spreads *T. brucei* among cattle, sheep, wild animals, and people. Symptoms develop within months to years of infection. Treatment is with antimicrobial drugs. No vaccine currently exists.

Primary Amebic Meningoencephalopathy

Two amoebae in the kingdom Euglenozoa, *Acanthamoeba* and *Naegleria*, cause **primary amebic meningoencephalopathy**, a rare disease with symptoms similar to those of other forms of meningitis and encephalitis. However, with primary amoebic meningoencephalopathy, the symptoms progressively worsen over a period of three to seven days until the patient dies. The amoebae live in warm bodies of water, mud, and moist soil, as well as in artificial water systems such as air-conditioning units and dialysis machines. They enter a host through the eyes or abrasions of the skin, or through inhalation of water while swimming, then migrate to the brain via cranial nerves. Combination antimicrobials have limited success, and only when administered very early in the infection: only three people have survived infection in the United States since records have been kept.

Prion Disease (pp. 622–624)

A **prion** is an infectious protein. In sheep, prions cause a disease called *scrapie.* In cattle, prions cause "mad cow disease." Both diseases are classified as *spongiform encephalopathies,* so called because they leave the infected animal's brain riddled with holes, like a sponge.

Variant Creutzfeldt–Jakob Disease (vCJD)

Creutzfeldt–Jakob disease is a naturally occurring dementia that strikes about one person in a million at approximately 60 years of age. It is not contagious. In contrast, **variant Creutzfeldt–Jakob disease (vCJD)** is contracted from eating the contaminated meat of animals infected with prions. In vCJD, brain tissue is progressively destroyed, causing patients to lose control of their muscles and experience insomnia, weight loss, and memory failure. Death usually occurs within 12 months of onset of symptoms. The incubation period is likely decades.

No treatment for vCJD is available; in fact, destruction of prions even outside the body is problematic. Prions survive cooking, freezing, pickling, and even normal autoclaving.

Microbial Diseases of the Eyes (pp. 624–625)

The senses are considered parts of the nervous system. Vision is our primary sense; in fact, the eyes can be considered extensions of the brain and dura mater.

Structure of an Eye

The *fibrous tunic* of the eye provides a tough barrier against microbes. It consists of the *sclera* and *cornea,* the latter of which covers the front of the eye. The *conjunctiva* lines the back side of the eyelids and all but the center of the cornea. The interior of the eye has two fluid-filled chambers called *humors.* The back inner wall of the eye is the *retina.* Its billions of sensory neurons respond to light by sending nerve impulses along the optic nerve to the brain.

Trachoma

Trachoma is the leading cause of nontraumatic blindness in humans. It develops in newborns when *Chlamydia trachomatis* present in a woman's reproductive tract enters into the baby's eyes during birth. Bacterial infection causes scarring of the conjunctiva and cornea. Corneal scarring triggers an invasion of blood vessels into the normally clear surface of the eye, resulting in blindness. Antimicrobials are administered for treatment of pregnant women. The eyes of infected newborns are treated with erythromycin cream.

Other Microbial Diseases of the Eyes

Conjunctivitis (pinkeye) is an inflammation of the conjunctiva due to infection with any of a variety of bacteria, fungi, viruses, or protozoa. **Keratitis** is inflammation of the cornea, and may also be caused by a variety of microbes. Treatment is with topical antimicrobial drugs; however, no antiviral drugs are available.

QUESTIONS FOR FURTHER REVIEW

Answers to these questions can be found in the answer section at the back of this study guide. Refer to the answers only after you have attempted to solve the questions on your own.

Multiple Choice

1. Which of the following is not a main component of the brain?
   a. cauda equina
   b. brain stem
   c. cerebellum
   d. cerebrum

2. Which of the following is a function of the cerebrospinal fluid?
   a. waste removal
   b. supply nutrients
   c. shock absorber
   d. All of the above are correct.

3. Motor neurons carry signals
   a. toward the CNS.
   b. away from the CNS to other organs.
   c. both toward and away from the CNS.
   d. None of the above is correct.

4. Inflammation of the brain is called
   a. meningitis.
   b. petechiae.
   c. encephalitis.
   d. otitis.

5. Ninety percent of bacterial meningitis cases are caused by five bacterial species. Which of the following is not one of those five species?
   a. *Haemophilus influenzae*
   b. *Streptococcus pneumonia*
   c. *Streptococcus agalactiae*
   d. *Staphylococcus aureus*

6. The leading cause of meningitis in adults is
   a. *Listeria monocytogenes.*
   b. *Streptococcus pneumonia.*
   c. *Streptococcus agalactiae.*
   d. *Staphylococcus aureus.*

7. Listeria avoids immune recognition by
   a. inhibiting immune cell functions.
   b. growing inside of macrophages.
   c. antigenic variation through mutation.
   d. None of the above is correct.

8. Mycolic acid in the cell wall of *Mycobacterium leprae* is responsible for all of the following except
   a. protection from antibody recognition.
   b. slow growth.
   c. resistance to antimicrobial drugs.
   d. protection from lysis after phagocytosis.

9. How could a person get botulism?
   a. contamination of wounds with spores
   b. ingestion of the toxin in canned foods
   c. ingestion of spores
   d. All of the above are correct.

10. Lockjaw is a common name for
    a. tetanus.
    b. botulism
    c. rabies.
    d. polio.

11. Tetanospasmin blocks the release of
    a. acetylcholine.
    b. inhibitory neurotransmitters.
    c. stimulatory neurotransmitters.
    d. None of the above is correct.

12. The most common form of meningitis is caused by
    a. protozoa.
    b. bacteria.
    c. fungi.
    d. viruses.

13. Poliovirus is most often acquired through
    a. eating contaminated food.
    b. respiratory secretions.
    c. drinking contaminated water.
    d. none of the above.

14. Most infections with poliovirus result in
    a. paralytic polio.
    b. asymptomatic polio.
    c. nonparalytic polio.
    d. minor polio.

15. Which of the following modes of transmission have been associated with rabies?
    a. animal bites
    b. contact with infected animals
    c. inhalation
    d. All the above are correct.

16. Arboviruses are carried among hosts by
    a. ticks.
    b. fleas.
    c. mosquitoes.
    d. flies.

17. Which of the following is not a disease caused by prions?
    a. trachoma
    b. Bovine spongiform encephalitis
    c. scrapie
    d. variant Creutzfeldt–Jakob disease

18. The structure of the eye responsible for responding to light energy is the
    a. conjunctiva.
    b. retina.
    c. cornea.
    d. fibrous tunic.

Fill in the Blanks

1. The two cell types of the nervous system are ____________________ and ____________________.

2. An environment that has no normal microbiota is called ____________________.

3. The nonprogressive form of leprosy is called ____________________.

4. ____________________ and ____________________ are the only natural hosts for *Mycobacterium leprae*.

5. Three different forms of botulism are ___________________, ___________________, and ___________________.

6. By binding to neuronal cytoplasmic membranes, botulism neurotoxins prevent the release of ___________________.

7. Most cases of viral meningitis are caused by viruses of the genus ___________________.

8. Botulism toxins interfere with neurotransmission to cause ___________________; tetanus toxins also interfere with neurotransmission, but instead cause muscles to continually ___________________.

9. The leading cause of nontraumatic blindness in humans is caused by the microorganism ___________________.

10. Arbovirus stands for ___________________.

11. *Trypanosoma brucei* is transmitted to its hosts via the ___________________.

12. The clear, colorless covering of the front of the eye is called the ___________________.

13. ___________________ is a bacterial disease in which blindness can result due to scarring of the conjunctiva and cornea.

14. Conjunctivitis is commonly known by the name ___________________.

Matching

Match the organism on the left to its proper description on the right. Each letter will be used only once.

1. ___ *Neisseria gonorrhoeae*	A. Trachoma
2. ___ *Naegleria* species	B. Bacterial meningitis
3. ___ *Mycobacterium leprae*	C. Fungal meningitis
4. ___ Coxsackie A virus	D. Ophthalmia neonatorum
5. ___ *Cryptococcus neoformans*	E. African sleeping sickness
6. ___ *Chlamydia trachomatis*	F. Primary amebic meningoencephalopathy
7. ___ *Streptococcus agalactiae*	G. Viral meningitis
8. ___ *Trypanosoma brucei*	H. Hansen's disease

Short-Answer Questions for Thought and Review

1. Briefly explain how signals are transmitted from an axon to a postsynaptic cell.

2. Because the CNS does not have any normal microbiota, how do microbes gain access to this region of the body?

3. Does infant botulism develop from the same cause as food-borne botulism?

4. Describe the differences in the four different conditions caused by poliovirus infection.

5. What is a prion, and how can abnormal ones be generated?

Critical Thinking

1. Why are intoxications by bacterial toxins often more deadly than the bacterial infections that produce the toxins?
2. Table 20.1 of the text compares the Salk and Sabin polio vaccines. Based on this information, information about viruses in general, and material found in the text regarding epidemiology and vaccines, discuss why the Sabin vaccine really shouldn't be used for eradication efforts. Why is the Salk vaccine a better choice for eradication even though more than one shot would be required?

Concept Building Questions

1. Most cases of bacterial meningitis are caused by just a few different bacterial species (the five discussed in the text). Discuss the virulence factors used by these bacteria that aid in their ability to cause disease, commenting on any trends seen among the different species.
2. There are a number of microorganisms that are able to evade the immune systems of their hosts. Two examples are *Listeria monocytogenes* and *Trypanosoma brucei*. Discuss how these two organisms use different methods to avoid the host immune response.

21 Microbial Cardiovascular and Systemic Diseases

CHAPTER SUMMARY

Structures of the Cardiovascular System (p. 632)

The heart, blood, and blood vessels comprise the *cardiovascular system,* which, along with the lymphatic system, transports fluids throughout the body. *Arteries* carry blood away from the heart to the lungs (via the *pulmonary arteries*) and to the rest of the body (via the *aorta*). They connect via *capillaries* to *veins,* which carry blood back to the heart from the lungs (via *pulmonary veins*) and from the rest of the body (via the *superior* and *inferior venae cavae*). Blood consists of *serum*, the liquid portion, and *formed elements* (*erythrocytes, leukocytes,* and *platelets*). *Plasma* is the liquid remaining when formed elements and clotting proteins are removed from blood.

Structure of the Heart

The heart is composed of two parallel pumping chambers, each of which contains an upper *atrium* and a lower *ventricle.* These are separated by valves. The heart tissue is composed of three layers: an outer fibrous *pericardium,* a muscular *myocardium,* and a thin inner *endocardium.*

Movement of Blood and Lymph

Blood flows through the cardiovascular system in the following sequence: venae cavae, right atrium, atrioventricular (tricuspid) valve, right ventricle, pulmonary semilunar valve, pulmonary arteries, lungs, pulmonary veins, left atrium, left atrioventricular (mitral) valve, left ventricle, aortic semilunar valve, aorta, arteries, capillaries, veins, venae cavae.

Blood is normally *axenic*; that is, it contains no microbes. Breaks in the skin or mucous membranes provide a route for microbes to enter the blood. The blood and lymph may spread these microbes throughout the body to cause **systemic diseases.**

Bacterial Cardiovascular and Systemic Diseases (pp. 632–646)

Bacteria can infect the blood as well as the blood vessels and heart.

Septicemia, Bacteremia, and Toxemia

In a patient with **septicemia**, pathogens are present in the blood and cause illness. When septicemia results in inflammation of the lymphatic vessels, **lymphangitis** results. **Bacteremia** refers specifically to bacterial septicemia; however, some physicians use the terms *bacteremia* and *septicemia* interchangeably. The cause of septicemia in the majority of patients is hidden and therefore it is called **occult septicemia.**

Septicemia is characterized by fever, chills, nausea, vomiting, shortness of breath, malaise, and changes in mental status. These signs and symptoms can progress rapidly to **septic shock,** a life-threatening condition of extremely low blood pressure resulting from dilation of blood vessels. Bacterial septicemia can also trigger minute skin lesions called **petechiae** on the trunk and lower extremities, or invade bones to cause **osteomyelitis,** inflammation of the bone and marrow.

When bacteria remain fixed at a site of infection, but release toxins into the blood, the condition is called **toxemia**. Living microbes release **exotoxins,** which can inhibit cell function or cause cell death. When dying Gram-negative bacteria disintegrate, they release **endotoxin**, which is the **lipid A** portion of **lipopolysaccharide (LPS),** from the outer layer of their cell membrane. Once released into the blood, endotoxin activates nearly all of the innate defensive reactions of the body, including widespread and severe coagulation (blood clotting), an often fatal condition called **disseminated intravascular coagulation (DIC)**. Endotoxin can also trigger macrophages to release potent cytokines, including **tumor necrosis factor** (TNF), **interleukins** (ILs), and **platelet activating factor** (PAF), which are important defenses in localized infections but life threatening when released throughout the body in septicemia. Treatment usually involves prompt administration of antimicrobial drugs.

Endocarditis

Bacterial colonization of the endocardium triggers an inflammatory condition called **endocarditis** and the formation of **vegetations,** masses of platelets and clotting proteins that bury the bacteria, hiding them from defensive cells, antibodies, and antimicrobial drugs. Fragments of vegetations and blood clots, called **emboli,** can lodge in small blood vessels, stopping the flow of blood and causing damage such as a stroke. Signs and symptoms of endocarditis include fever, fatigue, malaise, difficulty in breathing, tachycardia, and murmurs. These signs and symptoms may develop over a period of weeks or months (**subacute endocarditis**) or they may develop quickly (*acute endocarditis*). About half of all cases of endocarditis are caused by the *viridans streptococci,* but dozens of other bacteria can also cause the disease. Transmission is often via an infected tooth, dental trauma, or small skin wounds, contaminated catheters, syringes, or other medical devices. Diagnosis can be confirmed by visualization of vegetations via an **echocardiogram,** an image of the heart produced by sound waves. Treatment is with intravenous antibacterial drugs. Surgical removal of vegetations is requisite.

Brucellosis

Infection with *Brucella melitensis,* a small, nonmotile, aerobic Gram-negative coccobacillus, causes **brucellosis.** The classic symptom is a fluctuating fever that spikes every afternoon, giving the disease one of its common names, *undulant fever.* The patient also has chills, sweating, headache, myalgia, and weight loss. Humans become infected either by consuming unpasteurized contaminated dairy products or through contact with animal blood, urine, or placentas. The bacteria enter the body through breaks in the mucous membranes of the digestive or respiratory tracts and travel to the lymph nodes, liver, spleen, bone marrow, and heart. Physicians treat brucellosis with antibacterial drugs; however, most cases are mild and require no treatment. An attenuated vaccine for animals exists, but it is not administered to humans.

Tularemia

Another small, nonmotile, aerobic Gram-negative coccobacillus, *Francisella tularensis,* causes **tularemia,** a zoonotic disease transmissible to humans. The bacterium lives in animals and in blood-sucking ticks and insects, and is usually transmitted to humans via direct contact with an infected animal or a bite from an infected tick, mosquito, fly, or mite. It can also be transmitted via inhalation during animal slaughter, or via consumption of contaminated water or meat. Signs and symptoms are typical of many viral infections, and include fever, chills, headache, myalgia, and fatigue. Lymph nodes may be swollen. Treatment is with antimicrobial drugs. Though it is typically innocuous, tularemia is fatal to about 5% of untreated patients. An attenuated vaccine can lessen the severity of the disease and is administered to people at risk of exposure.

Plague

Plague (also called **bubonic plague** or the "black death") is caused by the bacillus *Yersinia pestis.* Its classic symptom is the sudden appearance of painfully inflamed lymph nodes called **buboes.** Other symptoms include fever, chills, headache, myalgia, septicemia, and blackening of the skin. **Pneumonic plague** occurs when the plague bacilli spread from the bloodstream to the lungs, or are inhaled.

Yersinia pestis carries virulence plasmids that code for **adhesions** and **type II secretion systems,** which inject antiphagocytic proteins preferentially into dendritic cells, neutrophils, or macrophages, neutralizing their ability to mount an adaptive immune response or eliminate the bacterium. Fleas spread the bacterium among rodents as well as to other animals and humans. Rapid treatment with antibacterial drugs is crucial. Untreated bubonic plague is fatal in 50% of cases, usually within one week of the onset of symptoms.

Lyme Disease

The bite of an *Ixodes* tick carrying the spirochete *Borrelia burgdorferi* causes **Lyme disease.** The initial symptom is a red "bull's-eye" rash at the site of the tick bite, as well as fever, headache, stiff neck, myalgia, and swollen lymph nodes. In about 10% of patients, the disease causes neurological symptoms (such as meningitis) and cardiac dysfunction. About 80% of patients develop severe arthritis as a result of the body's immunological response to the bacterium. The disease is common in woodland areas populated by deer, since the *Ixodes* tick feeds on deer. Treatment is with antimicrobial and anti-inflammatory drugs. No vaccine is currently available.

Erlichiosis and Anaplasmosis

The Gram-negative bacteria *Erlichia chaffeensis* and *Anaplasma phagocytophilum* can be transmitted to humans by hard-shelled ticks, causing **erlichiosis** and **anaplasmosis**, respectively. Both illnesses have flulike symptoms and can cause a decrease in white blood cells (*leukopenia*) and platelets (*thrombocytopenia*) in the blood. The symptoms can be life threatening, depending on the patient's age and health. Antibiotics should be administered immediately to limit complications and mortality. No vaccine is available for either organism.

Viral Cardiovascular and Systemic Diseases (pp. 646–652)

Many viral diseases spread throughout the body via the cardiovascular and lymphatic systems.

Infectious Mononucleosis

Infectious mononucleosis (*kissing disease*) results from infection with *human herpesvirus 4 (HHV-4),* which is also known as *Epstein-Barr virus (EBV).* This enveloped, dsDNA virus suppresses **apoptosis** (programmed cell death) of B lymphocytes, causing infected cells to become immortal. Such infected B cells are one source of cancers such as **Burkitt's lymphoma** when certain cofactors are present. EBV is usually transmitted in saliva. When EBV enters the blood and invade B lymphocytes, they trigger long-term immune responses, which causes the symptoms and signs of mononucleosis. No specific treatment exists; care involves relief of symptoms. Most patients recover within four weeks.

Cytomegalovirus Disease

Another herpesvirus that affects humans is *Cytomegalovirus (CMV),* an enveloped, dsDNA virus that causes infected cells to become enlarged. CMV typically remains latent unless the immune system is compromised; however, in fetuses, newborns, and immunocompromised patients, infection can cause an enlarged liver and spleen, jaundice, anemia, deafness, blindness, pneumonia, and even death. Transmission is typically via sexual intercourse, *in utero* exposure, vaginal birth, blood transfusions, or organ transplants. Treatment of newborns is difficult because in most cases the damage is irreversible prior to diagnosis. In adults, the drug *fomiversin* is injected into the eye to inhibit replication of CMV in retinal cells. There is no vaccine against CMV.

Yellow Fever

Yellow fever is caused by the virus *Flavivirus,* an enveloped +ssRNA virus carried by *Aedes aegypti* mosquitoes, which feed on humans. After a bite from an infected mosquito, the virus travels to the liver. Signs and symptoms develop within three to six days and may follow three stages: The earliest stage is a few days of mild fever, headache, muscle aches, and vomiting. This is followed by a period of remission. Fifteen percent of patients proceed to a third stage characterized by delirium, seizures, coma, and degeneration of the liver, kidneys, and heart, as well as hemorrhages and shock. The accompanying jaundice gives the disease its name. There is no specific treatment. About 20% of patients with the severe form die.

Dengue Fever and Dengue Hemorrhagic Fever

Dengue viruses are enveloped +ssRNA viruses in the genus *Flavivirus.* They are carried by *Aedes* mosquitoes, and infect millions of people worldwide each year. The bite of an infected mosquito causes **dengue fever** (or *breakbone fever*), which manifests in fever, weakness, edema, and severe pain. Dengue fever is self-limiting, lasting six to seven days. Reinfections result in a much more severe disease: **dengue hemorrhagic fever (DHF)** is a serious hyperimmune response in which activated memory T cells release inflammatory lymphokines, triggering rupture of blood vessels, internal bleeding, shock, and possibly death. No specific treatment is available.

African Viral Hemorrhagic Fevers

Ebola hemorrhagic fever, which is caused by *Ebolavirus,* and **Marburg hemorrhagic fever,** which is caused by *Marburgvirus,* are endemic to Africa and are emerging global health concerns. Both begin with fever, fatigue, dizziness, muscle pain, and exhaustion, followed by minor petechiae. This progresses to severe bleeding internally and from other body orifices, including the mouth, eyes, and ears. Death results from shock, seizures, or kidney failure. Ebolavirus and Marburgvirus are –ssRNA viruses that are thought to initially infect humans via a bat host. Humans transmit the viruses via contact with bodily fluids. There are no effective antiviral drugs to treat hemorrhagic fevers. Ebolavirus is fatal in up to 90% of patients, and Marburgvirus in 25%.

Protozoan and Helminthic Cardiovascular and Systemic Diseases (pp. 653–661)

Of the protozoan and helminthic cardiovascular and systemic diseases, the most prevalent and serious is malaria.

Malaria

Apicomplexans are protozoa whose infective forms have an ornate complex of organelles at their apical ends. **Malaria** is caused by apicomplexans of the genus *Plasmodium,* four species of which infect the liver and erythrocytes, causing recurrent fever and chills, anemia, weakness, and fatigue. The inability of the liver to process the hemoglobin released from so many dying erythrocytes causes severe jaundice. Worldwide, two to three million people die each year from malaria. If a victim survives the acute stages, immunity gradually develops and periodic episodes become less severe.

The life cycle of *Plasmodium* has three stages: (1) In the **exoerythrocytic cycle,** *Plasmodium* sporozoites injected by a female *Anopheles* mosquito infect the host's liver. (2) In the **erythrocytic cycle,** merozoites penetrate erythrocytes and become trophozoites, which cause cyclical lysis of erythrocytes. Some merozoites develop into male and female gametocytes. (3) In the **sporogonic cycle,** a female mosquito feeding on an infected human ingests gametocytes within erythrocytes. Eventually, male gametes fertilize female gametes to produce zygotes, which develop into sporozoites that can then be transmitted to a new host.

Malaria is treated with the drugs chloroquine and pyrimethamine, or with combination regimens in resistant areas. Prophylactic medication can prevent infection for travelers into endemic areas. Researchers are working to develop antimalarial vaccines.

Toxoplasmosis

The apicomplexan protozoan *Toxoplasma gondii* causes **toxoplasmosis.** The majority of infected people have no symptoms; however, transplacental transfer from mother to fetus can cause a variety of birth defects or fetal demise. Infection in immunocompromised patients can result in paralysis, blindness, myocarditis, encephalitis, and death. Cats are the definitive host for *T. gondii,* and shed immature oocysts in their feces. Humans typically become infected by consuming undercooked meat containing the parasite. When treatment is recommended—such as in AIDS patients, pregnant women, and newborns—combination antimicrobials are used for three to four weeks. A vaccine for cats is currently under development to reduce the chance of pet-to-owner transmission.

Chagas' Disease

The flagellated protozoan *Trypanosoma cruzi* causes **Chagas' disease,** which is endemic throughout Central and South America. Initial symptoms, which are experienced by only 1% of infected people, are swelling at the site of infection, followed by fever, malaise, and swollen lymph nodes for four to eight weeks. The disease then enters an asymptomatic stage for 10–20 years. In some patients, this is followed by a symptomatic stage characterized by fatal congestive heart failure following the formation of *pseudocysts,* which are clusters of the parasite in heart muscle tissue.

Transmission occurs via the bite of blood-sucking bugs in the genus *Triatoma. T. cruzi* matures in the bug's hindgut and enters a host when the bug's feces are rubbed into a bite wound. It circulates in the blood before infecting macrophages and heart muscle cells. Chagas' disease can be diagnosed by **xenodiagnosis,** a procedure in which a physician allows an uninfected bug to feed on a person suspected of having the disease and then seeks the trypanosome in the bug's hindgut. Most patients do not take antitrypanosome drugs soon enough; late stages of Chagas' disease cannot be treated. No vaccine exists.

Schistosomiasis

Parasitic blood flukes of the genus *Schistosoma* are among the most common parasitic helminths in the world. They cause **schistosomiasis,** a potentially fatal disease that manifests initially as a transient dermatitis called **swimmer's itch** where infective larvae burrow into the skin. The larvae migrate through the vascular system, then mature into worms that mate to produce eggs. These lodge in the liver, brain, and other organs, causing organ failure and sometimes death. Humans become infected by contact with contaminated water. Snails are the intermediate host. The drug of choice is praziquantel. A vaccine is currently in clinical trials.

QUESTIONS FOR FURTHER REVIEW

Answers to these questions can be found in the answer section at the back of this study guide. Refer to the answers only after you have attempted to solve the questions on your own.

Multiple Choice

1. Which of the following is not a formed element found in the blood?
   a. platelets
   b. clotting proteins
   c. erythrocytes
   d. leukocytes
2. What is the outermost layer of the heart called?
   a. myocardium
   b. endocardium
   c. pericardium
   d. None of the above is correct.
3. Nutrients and oxygen diffuse into tissue through
   a. arteries.
   b. veins.
   c. interstitial spaces.
   d. capillaries.
4. Lipid A is also referred to as
   a. endotoxin.
   b. LPS.
   c. exotoxin.
   d. cytotoxin.
5. Endotoxin activates which of the following host defenses?
   a. blood clotting
   b. complement
   c. inflammation
   d. All the above are correct.

6. Half of the cases of bacterial endocarditis are caused by
   a. *Neisseria.*
   b. Viridans streptococci.
   c. *Mycoplasma.*
   d. *Staphylococcus epidermidis.*
7. Interruption of the blood supply to a tissue is called
   a. ischemia.
   b. endocarditis.
   c. necrosis.
   d. lymphangitis.
8. Undulant fever is also referred to as
   a. brucellosis.
   b. gas gangrene.
   c. tularemia.
   d. ischemia
9. Which of the following can transmit *Francisella* to humans?
   a. ticks
   b. mosquitoes
   c. flies
   d. All the above are correct.
10. Pneumonic plague occurs when bacteria infect the
    a. lymph nodes.
    b. lungs.
    c. heart.
    d. None of the above is correct.
11. *Yersinia pestis* is spread among rodents by what vector?
    a. ticks
    b. flies
    c. fleas
    d. mosquitoes
12. The final stage of Lyme disease is characterized by
    a. arthritis.
    b. meningitis.
    c. encephalitis.
    d. a "bull's-eye" rash.
13. *Borrelia burgdorferi* uses which of the following in its metabolic processes?
    a. zinc
    b. iron
    c. copper
    d. manganese
14. Burkitt's lymphoma is caused by
    a. *Cytomegalovirus.*
    b. *Marburgvirus.*
    c. Epstein-Barr virus.
    d. herpes simplex virus.
15. What is considered the most significant viral disease of humans?
    a. infectious mononucleosis
    b. yellow fever
    c. cytomegalovirus disease
    d. Ebola hemorrhagic fever
16. Which *Plasmodium* species causes the most severe form of disease?
    a. *P. ovale*
    b. *P. falciparum*
    c. *P. malariae*
    d. *P. vivax*
17. Diagnosis of schistosomiasis is made by identifying which of the following in stool or urine?
    a. eggs
    b. cercariae
    c. miracidia
    d. None of the above is correct.

Fill in the Blanks

1. The heart, blood, and blood vessels make up the ____________________.
2. Bacterial invasion of the bones is called ____________________.
3. ____________________ is the condition of extremely low blood pressure resulting from dilation of blood vessels.

4. A(n) __________________ infection is one acquired in a health care setting.

5. Septicemia in which the causative bacteria can't be cultured is called

__________________.

6. Penicillin is ineffective against *F. tularensis* because it produces

__________________.

7. Inflamed lymph nodes that are one manifestation of plague are called

__________________.

8. __________________ is the most reported vector-borne disease in the United States.

9. __________________ is a decrease in white blood cells in the blood.

10. Viral hemorrhagic fevers occur mainly in __________________.

11. __________________ are the reservoir for yellow fever virus.

12. __________________ is the form of asexual reproduction seen in the apicomplexan life cycle.

13. __________________ are the definitive hosts of *Toxoplasma gondii.*

14. __________________ are groups of parasites in the heart muscle tissue of a patient with Chagas' disease.

Matching

Match the organism on the left to its corresponding disease on the right. Each letter will be used only once.

1. ____ Viridans streptococci
2. ____ *Pseudomonas aeruginosa*
3. ____ *Trypanosoma cruzi*
4. ____ Epstein-Barr virus
5. ____ *Escherichia coli*
6. ____ *Borrelia burgdorferi*
7. ____ *Francisella tularensis*
8. ____ Ebolavirus
9. ____ *Plasmodium vivax*
10. ____ *Yersinia pestis*

A. Septicemia
B. Viral hemorrhagic fever
C. Endocarditis
D. Tularemia
E. Malaria
F. Plague
G. Chagas' disease
H. Bacteremia
I. Lyme disease
J. Infectious mononucleosis

Short-Answer Questions for Thought and Review

1. Briefly discuss the traits that provide some individuals with resistance to malaria.

2. How does *Borrelia burgdorferi* get around the problem of limiting iron concentrations in hosts and the environment? Why is this significant?

3. Describe the life cycle of *Schistosoma*.

Critical Thinking

1. At what points during the malarial life cycle would antibodies be able to recognize *Plasmodium* parasites? If you attempted to passively immunize someone with antimalarial antibodies, would this stop the infection?
2. Discuss why dengue hemorrhagic fever is so much worse than dengue fever.
3. Diagnosis of many helminthic diseases relies on microscopic identification of eggs in stool samples. How reliable is this method? Would you use this method? Why, in general, aren't there better methods of detection and diagnosis?

Concept Building Questions

1. Explain the immunological battle that leads to the production of symptoms in infectious mononucleosis. In someone with immunosuppression, would you expect to see symptoms? What about in someone with hypersensitivities?
2. Contrast the infectiousness for humans of plague if it undergoes an amplification cycle in animals versus if it does not. Include in your answer what an amplification cycle is and why humans would be more or less at risk of disease depending on whether amplification occurs.

22 Microbial Diseases of the Respiratory System

CHAPTER SUMMARY

Structures of the Respiratory System (pp. 668–670)

The respiratory system functions to exchange gases between the atmosphere and the blood. Anatomists commonly divide its structures into upper and lower systems.

Structures of the Upper Respiratory System, Sinuses, and Ears

The upper respiratory system includes the nose, the nasal cavity, and the pharynx. These structures are lined with tiny hairs and a ciliated mucous membrane that filter inspired air. Sinuses carry fluids and infecting microorganisms into the nasal cavity, and ducts from the eyes carry contaminants into the pharynx. Auditory tubes allow the spread of contaminants from the pharynx to the ears. Lymphoid tissues called *tonsils* contain cells and chemicals to combat microbes in the nasal cavity, pharynx, and auditory tubes. The mucus of the upper respiratory system contains antimicrobial chemicals.

Structures of the Lower Respiratory System

The lower respiratory system begins at the larynx, which contains the vocal cords, and continues as the trachea (commonly called the *windpipe*). Air entering the trachea flows into a succession of smaller branches, from the bronchi to the bronchioles to the alveoli, where gas exchange occurs.

The synchronous beating of cilia lining the trachea, bronchi, and bronchioles carries mucus and trapped contaminants up to the pharynx in an action referred to as the **ciliary escalator**. Further defense against microorganisms is provided by alveolar macrophages and secretory antibodies present in tears, saliva, and respiratory mucus. Successful invasion by pathogens results in inflammation of the respiratory tract, which in turn restricts airflow.

Normal Microbiota of the Respiratory System

The lower respiratory system normally lacks microorganisms; however, the upper respiratory system is inhabited by many types of microbes. These include harmless **diphtheroids** in the nose and nasal cavity, as well as opportunistic *Staphylococcus* and *Streptococcus* species in the pharynx.

Bacterial Diseases of the Upper Respiratory System, Sinuses, and Ears (pp. 670–674)

Bacteria can infect the upper respiratory system and spread into the sinuses and auditory tubes.

Streptococcal Respiratory Diseases

Physicians recognize a variety of diseases of the upper respiratory system caused by species of *Streptococcus*. These include **streptococcal pharyngitis** (strep throat), an inflammation of the pharynx in which purulent abscesses cover the tonsils. If the bacteria spread into the lower respiratory tract, they may cause **laryngitis** (inflammation of the larynx resulting in hoarseness) or **bronchitis** (inflammation of the bronchi resulting in labored breathing and coughing). Streptococcal pharyngitis can also progress to **scarlet fever** (scarlatina) when the microbe releases toxins that trigger fever and a diffuse rash, followed by skin sloughing. Untreated streptococcal pharyngitis can spread to cause infection of the kidneys (*acute glomerulonephritis*) and **rheumatic fever**, inflammation leading to damage of the heart valves and muscle.

Lancefield group A streptococcus (also known as *S. pyogenes*) is the major cause of bacterial pharyngitis and scarlet and rheumatic fevers. The species' virulence factors include: M proteins that destabilize a component of complement; a hyaluronic capsule that may provide "camouflage"; streptokinase and other enzymes that assist the microbe's dissemination; pyrogenic toxins; and streptolysins that lyse blood cells and platelets. The microbe travels via respiratory droplets. Penicillin provides effective treatment.

Diphtheria

Diphtheria, a potentially fatal respiratory disease, is endemic in countries that lack adequate immunization. Symptoms include sore throat, fever, and the formation of a pseudomembrane that adheres to the tonsils, uvula, roof of the mouth, pharynx, and larynx and can cause suffocation. *Corynebacterium diphtheriae*, a Gram-positive bacterium, produces the potent *diphtheria toxin* that causes the disease. Timely administration of antitoxin is essential; destruction of *Corynebacterium* with penicillin or erythromycin prevents the synthesis of more toxin. Severe cases require surgery to open the blocked airway, or insertion of a tracheostomy tube. Immunization during childhood must be followed by booster immunizations every 10 years.

Sinusitis and Otitis Media

Sinusitis (inflammation of the sinuses) and **otitis media** (inflammation of the middle ear) are usually caused by *S. pneumoniae, Haemophilus influenzae,* or *Moraxella catarrhalis,* which spread from the pharynx to the sinuses or ears. Sinusitis typically causes pain and pressure in the affected region. Otitis media manifests in severe pain in the ear, which ends abruptly when the tympanic membrane ruptures, releasing the pressure. Children are much more likely to suffer otitis media than adults, because their auditory tubes are smaller and more horizontal, facilitating accumulation of fluid.

Viral Diseases of the Upper Respiratory System (pp. 674–675)

Viral diseases of the upper respiratory system are among the most common human diseases.

Common Cold

Of the over 200 different serotypes of viruses that cause common cold, the major culprits are in the genus *Enterovirus* (family *Picornaviridae*). Some coronaviruses, adenoviruses, reoviruses, and paramyxoviruses also cause common cold. Symptoms of common cold are sore throat, sneezing, rhinorrhea, cough, and congestion. Fever indicates bacterial infection. Cold viruses reproduce most effectively at about 33°C, which is the temperature of the nasal cavity. The higher temperatures of the lower respiratory tract inhibit or destroy them. A single rhinovirus is sufficient to cause a cold in 50% of infected individuals. When symptoms are most severe, there may be over 100,000 virions/ml of nasal mucus, and these virions remain infective for hours outside the body. An effective vaccine is not practical, as it would have to protect against all the viral serotypes. Antiseptic measures such as frequent handwashing are the most important preventive measures.

Bacterial Diseases of the Lower Respiratory System (pp. 675–685)

The lower respiratory organs are axenic; that is, they are normally devoid of microorganisms. Successful bacterial colonization can cause life-threatening disease.

Bacterial Pneumonias

The term **pneumonia** describes an inflammation of the lungs in which the alveoli and bronchioles become filled with fluid. Pneumonias are classified according to the affected region and the causative organism. The most common types of bacterial pneumonia include the following:

- **Pneumococcal pneumonia** is the most common type, accounting for most hospitalizations due to pneumonia. It is caused by *Streptococcus pneumoniae* (commonly known as **pneumococcus**) and is usually lobar. Blood frequently enters the lungs, causing rust-colored sputum. Penicillin is effective against most strains, though some are now resistant. A vaccine is available.
- **Primary atypical pneumonia** (also called *walking pneumonia*) is caused by *Mycoplasma pneumoniae*. It is the leading type of pneumonia in children and young adults. Early symptoms include fever, headache, sore throat, and excessive sweating. These are followed by the development of a persistent cough. Treatment is with erythromycin or tetracycline. No vaccine is available.
- ***Klebsiella* pneumonia** is caused by infection with the Gram-negative *Klebsiella pneumoniae*. Following inhalation, often in health care settings, it destroys alveolar cells, producing thick, bloody sputum and recurrent chills. Bacterial cell death releases endotoxin that can trigger shock and disseminated intravascular coagulation. Cephalosporin and aminoglycosides can be used against *Klebsiella*, but many strains are resistant, and the release of endotoxin can cause irreversible damage to the lungs. No vaccine is available.
- **Pneumonic plague** is a form of pneumonia caused by *Yersinia pestis*, the bubonic plague bacterium. Initial infection can progress to clinical pneumonia in just a few hours, and rapid shock and death result if the patient is not treated.
- **Ornithosis** in birds infected with *Chlamydophila psittaci* causes severe pneumonia in humans.
- *Chlamydophila pneumoniae* causes bronchitis, pneumonia, and sinusitis, but the infections are mild and usually do not require treatment.

Legionellosis (Legionnaires' Disease)

Legionellosis (or **Legionnaires' disease**) is named for an outbreak of the disease at an American Legion convention in 1976 during which 29 people died. Infection initially manifests as fever, chills, cough, and headache. If left untreated, rapid and severe inflammation of the lungs follows, resulting in death of up to 50% of patients. Nineteen species of *Legionella* are pathogenic, but *L. pneumophila* causes 85% of infections in humans. The bacterium invades freshwater protozoa, which release *Legionella*-filled vesicles into the environment via air-conditioning systems, vaporizers, hot tubs, and other devices. Quinolones or macrolides are the drug of choice for treatment.

Tuberculosis

Tuberculosis (**TB**) is an infection of the lungs caused by *Mycobacterium tuberculosis*. The symptoms are initially mild fever and cough, but progression of the disease leads to labored breathing, wheezing, chest pain, fatigue, weight loss, and coughing up blood. The cell walls of *M. tuberculosis* contain **mycolic acid,** a waxy lipid that protects the pathogen from lysis, dessication, and many antimicrobial drugs, and **cord factor**, which inhibits migration of neutrophils and kills host cells. In primary TB, infection remains in the lungs and causes the formation of small, hard nodules called tubercles that are filled with infected cells. With a healthy immune response, the disease becomes dormant at this point, and may remain so for decades. Reactivated TB can lead to dissemination to the kidneys, bone marrow, spinal cord, and brain. The common name for TB until the early 1900s—*consumption*—reflects the wasting resulting from involvement of multiple sites.

A tuberculin skin test determines whether a person has been exposed to *M. tuberculosis* antigens, but cannot determine whether a person has clinical disease. The presence of acid-fast cells and cords in sputum confirms active disease. Treatment requires a combination of drugs for six months. Multidrug-resistant strains have arisen in several countries. Vaccination induces immunity in 80% of patients and produces a "false positive" skin test for the rest of the person's life, hampering epidemiologic efforts.

Pertussis (Whooping Cough)

Infection with the Gram-negative coccobacillus *Bordetella pertussis* causes **pertussis** (**whooping cough**), a disease of children that is characterized by coughing so severe that it can cause vomiting, ruptured blood vessels in the eyes, choking, suffocation, and even death. Pertussis progresses in four stages: During **incubation**, the disease is asymptomatic. In the **catarrhal phase**, which lasts one to two weeks, the child manifests symptoms similar to those of the common cold. The characteristic whooping cough marks the **paroxysmal phase**, when bacterial toxins and adhesions disrupt ciliated epithelial cells of the trachea, stopping the ciliary escalator. In repeated attempts to clear the lungs of mucus, the child coughs without inhalation, then finally succeeds in inhaling through a congested trachea, producing a whooping sound. This phase lasts two to four weeks, and resolves in the **convalescent phase**, which continues for several more weeks as the cough subsides. Antimicrobial drugs have little effect, and treatment is largely supportive. An effective vaccine has been available in the United States since 1949.

Inhalation Anthrax

Bacillus anthracis causes the most lethal form of anthrax, **inhalation anthrax,** which progresses from symptoms of a common cold to lethargy, shock, and death within days unless aggressively treated. Anthrax spreads via inhalation of bacterial endospores in dust or on animal hides or wool. Many antimicrobials are effective against *B. anthracis,* but damage to the lungs and toxemia can be so severe by the time of diagnosis that death is common. An effective vaccine is available.

Viral Diseases of the Lower Respiratory System (pp. 685–693)

Of the viral diseases of the lower respiratory system, the most prevalent is influenza.

Influenza

The signs and symptoms of **influenza** (or **flu**) usually include sudden fever, pharyngitis, congestion, dry cough, malaise, headache, and myalgia. Most people recover within two weeks. Causative organisms include two species of orthomyxoviruses, designated types A and B. Orthomyxoviruses are surrounded by lipid envelopes with glycoprotein spikes composed of **hemagglutinin** (HA) or **neuraminidase** (NA) that help the viruses attach to the cells of the lungs. Mutations in the genes coding for these glycoprotein spikes are responsible for the production of new strains of influenzavirus. **Antigenic drift** refers to the accumulation of hemagglutinin and neuraminidase gene mutations in a geographic area, whereas **antigenic shift** is a major antigenic change that results from the reassortment of genes from different influenza A viruses infecting the same cell.

Annually, 200,000 Americans are hospitalized with influenza, and about 30,000 die. Antiviral therapy is effective only if taken in the first 48 hours of infection. Immunization with a multivalent vaccine is 70% effective against the antigens it contains. Immunity lasts for three years or less.

Severe Acute Respiratory Syndrome

Severe acute respiratory syndrome (SARS) is a newly emerging disease caused by a coronavirus that destroys lung cells and spreads via the bloodstream to the heart and kidneys. SARS manifests as a high fever, labored breathing, malaise, and body aches. No antiviral drug is effective against SARS, and in 2003, 774 people died from the condition worldwide.

Respiratory Syncytial Virus Infection

Respiratory syncytial virus (RSV) infection is the leading respiratory killer of infants worldwide. In infants and the immunocompromised, RSV triggers fever, rhinorrhea, coughing, and wheezing, as well as **bronchiolitis** (inflammation of the bronchioles) and pneumonia. Some children develop **tracheobronchitis** (inflammation of the trachea and bronchi), commonly known as **croup.** RSV is an enveloped –ssRNA virus that causes the formation of **syncytia**—giant, multinucleated cells formed from the fusion of virally infected cells to neighboring cells—in the lungs. Supportive care is given in mild cases. Severe cases require the administration of immunoglobulins against RSV. Handwashing is an important preventive measure. No vaccine is currently available.

Hantavirus Pulmonary Syndrome

Hantavirus is a genus of enveloped, –ssRNA viruses that infect species of mice. Infection with *Hantavirus* causes **hantavirus pulmonary syndrome** (**HPS**), a disease initially characterized by fever, fatigue, and muscle aches in the trunk and legs. These symptoms rapidly progress to coughing, difficulty in breathing, and shock, as inflammation causes the lungs to fill with fluid and the blood pressure to drop precipitously. Half of all patients do not survive. Transmission is via inhalation of dried mouse urine, feces, or saliva. No specific treatment is available, and no vaccine exists.

Other Viral Respiratory Diseases

Three strains of **parainfluenzaviruses** cause croup and **viral pneumonia**, particularly in young children. Most children recover from viral pneumonia with supportive care within a few days. **Metapneumovirus** (**MPV**) is the second most common cause of respiratory disease in children, and researchers have found that antibodies against the virus form in all children by age five.

Mycoses of the Lower Respiratory System (pp. 693–698)

Systemic mycoses are fungal infections that spread throughout the body.

Coccidioidomycosis

Also known as valley fever, **coccidioidomycosis** is caused by infection with *Coccidioides immitis*. Initial symptoms resemble those of pneumonia or TB, but can range from mild and self-limiting to severe. In the immunocompromised, dissemination to the central nervous system can result in meningitis, nausea, and emotional disturbance. CNS dissemination is fatal if untreated. Transmission is via inhalation of asexual spores called *arthroconidia* from disturbed soil. Treatment is with antifungal drugs.

Blastomycosis

Infection with *Blastomyces dermatitidis* causes **blastomycosis**, a fungal disease that begins with flulike symptoms. Dissemination of the disease may cause lesions on the skin or necrosis and cavity formation in bones and other tissues. Inhalation of dust carrying fungal spores infects the lungs, where spores germinate to form yeasts. Administration of antifungal drugs for 10 weeks or longer is required. Relapse is common. Mortality is high among immunocompromised patients.

Histoplasmosis

The most common fungal systemic disease affecting humans is **histoplasmosis**. It is asymptomatic in 95% of patients, but 5% of patients develop clinical disease, which can take four forms: pulmonary histoplasmosis causes severe coughing and bloody sputum; cutaneous histoplasmosis causes skin lesions; ocular histoplasmosis causes inflammation of the eyes; and systemic histoplasmosis causes a potentially fatal enlargement of the spleen and liver. The infection is caused by *Histoplasma capsulatum*, a dimorphic ascomycete that is found in moist soils. Treatment is with antifungal drugs.

Pneumocystis Pneumonia

Infection with *Pneumocystis jiroveci* causes ***Pneumocystis* pneumonia (PCP)** in malnourished, premature infants, weak elderly patients, and people with AIDS. Asymptomatic in healthy people, signs and symptoms in the immunocompromised include difficulty breathing, mild anemia, hypoxia, and fever. If left untreated, PCP involves more and more lung tissue until death occurs. Transmission is thought to occur via inhalation of droplet nuclei containing the fungus. Treatment is with a combination of trimethoprim and sulfamethoxazole.

QUESTIONS FOR FURTHER REVIEW

Answers to these questions can be found in the answer section at the back of this study guide. Refer to the answers only after you have attempted to solve the questions on your own.

Multiple Choice

1. Otitis media is most common in
   a. both adults and children.
   b. adults.
   c. children.
   d. It rarely occurs in either group.
2. The genus responsible for most colds is
   a. *Reovirus.*
   b. *Adenovirus.*
   c. *Coronavirus.*
   d. *Enterovirus.*
3. Colds can be spread by which of the following routes?
   a. coughing and sneezing
   b. hand-to-hand contact
   c. fomites
   d. All the above are correct.
4. The most common bacterial pneumonia is
   a. pneumococcal.
   b. mycoplasmal.
   c. pneumonic plague.
   d. *Klebsiella* pneumonia.
5. The main environmental habitat of *Legionella* is
   a. humans.
   b. water.
   c. soil.
   d. none of the above.
6. Tubercles are formed during
   a. disseminated tuberculosis.
   b. secondary tuberculosis.
   c. primary tuberculosis.
   d. all of the above.
7. Mycolic acid in the cell wall of *Mycobacterium* is responsible for all of the following characteristics except
   a. inhibition of neutrophil migration.
   b. slow growth.
   c. resistance to antimicrobial drugs.
   d. intracellular growth.
8. The lower respiratory system includes all of the following except
   a. trachea.
   b. bronchi.
   c. pharynx.
   d. larynx.
9. Which of the following contributes to the lack of microorganisms in the lower respiratory system?
   a. phagocytic cells
   b. IgA antibodies
   c. ciliary escalator
   d. All the above are correct.

10. Which of the following is not a complication seen with streptococcal pharyngitis?
    a. rheumatic fever
    b. whooping cough
    c. acute glomerulonephritis
    d. scarlet fever

11. The presence of a pseudomembrane is characteristic of what disease?
    a. whooping cough
    b. pneumonia
    c. diphtheria
    d. inhalation anthrax

12. Which of the following is not a way to transmit anthrax?
    a. inhalation of endospores
    b. person-to-person contact
    c. contact with infected animals
    d. Anthrax can be transmitted by all of these methods.

13. Which of the options below correctly notates the following strain of influenzavirus: influenzavirus type A containing both type 1 HA and NA antigens that was identified in the USSR in January 1978?
    a. B/1/78/USSR (H1N1)
    b. B/H1N1/USSR/January/78
    c. USSR/1/78/B (H1N1)
    d. B/USSR/1/78 (H1N1)

14. SARS was first identified in
    a. the United States.
    b. Japan.
    c. Canada.
    d. China.

15. The most common respiratory disease in newborns is
    a. respiratory syncytial virus infection.
    b. cytomegalovirus disease.
    c. influenza.
    d. mycoplasmal pneumonia.

16. Hantavirus pulmonary syndrome occurs from inhalation of virus in feces or urine from
    a. deer.
    b. mice.
    c. birds.
    d. none of the above.

17. *Coccidioides* is found most often in which of the following regions?
    a. the United States
    b. Canada
    c. China
    d. None of the above is correct.

18. The most common systemic fungal infection of humans is
    a. blastomycosis.
    b. pneumocystis pneumonia.
    c. histoplasmosis.
    d. coccidioidomycosis.

Fill in the Blanks

1. The main components of the upper respiratory system are the

   ____________________, ____________________, and ____________________.

2. ____________________ is produced by group A streptococci and allows

   the bacteria to digest blood clots.

3. In the lungs, ____________________ diffuses out of the alveoli, while

   ____________________ diffuses into the alveoli.

4. Infection of the middle ear by bacteria in the pharynx results in the condition known as ____________________.

5. ____________________ results in inflammation of the lungs with fluid accumulation in the alveoli and bronchioles.

6. ____________________ is the leading type of pneumonia in children.

7. The ____________________ is used to screen patients for exposure to *M. tuberculosis*.

8. ____________________ is the protein responsible for the ability of influenzavirus to bind pulmonary epithelial cells.

9. Giant, multinucleated cells formed by the fusion of virally infected cells with neighboring uninfected cells are called ____________________.

10. It is likely that ____________________ ranks second to rhinoviruses as a cause of viral respiratory disease.

11. Blastomycosis is mainly treated with the antifungal agent ________________.

12. ____________________ is a common fungal respiratory illness of persons with AIDS.

13. The characteristic cough of pertussis is seen during the ____________________ stage.

Matching

Match the organism on the left to its corresponding disease on the right. Each letter will be used only once.

1. ____ *Legionella pneumophila*
2. ____ *Bordetella pertussis*
3. ____ *Streptococcus pyogenes*
4. ____ Coronavirus
5. ____ Rhinovirus
6. ____ *Bacillus anthracis*
7. ____ *Mycoplasma pneumoniae*
8. ____ *Yersinia pestis*

A. Common cold
B. Severe acute respiratory syndrome
C. Walking pneumonia
D. Inhalation anthrax
E. Pneumonic plague
F. Scarlet fever
G. Pontiac fever
H. Whooping cough

Short-Answer Questions for Thought and Review

1. How does *Streptococcus pneumoniae* hide itself from the immune system?

2. Explain how diphtheria toxin contributes to the symptoms seen in an individual with diphtheria.

3. Individuals with a common cold sometimes complain about having the flu. Identify symptoms that could help someone distinguish the flu from a cold.

Critical Thinking

1. Rhinoviruses are ubiquitous and yet they are limited to causing upper respiratory tract infections that are rarely life-threatening. Why are they limited to this particular body region and why does this limit the severity of infection?
2. Why does immunosuppression increase the likelihood of infection and reactivation of tuberculosis?
3. Discuss the difference between the concepts of antigenic shift and antigenic drift that result in the constantly changing strains of influenzavirus. Would you expect new strains of influenzavirus type B to be generated by antigenic drift, antigenic shift, or both?

Concept Building Questions

1. *Legionella pneumophila* is a rather fastidious microorganism and yet it is able to survive in numerous aqueous environments. Discuss what factors contribute to the microorganism's widespread occurrence in these environments.
2. Explain why a positive tuberculin skin test alone is not sufficient to distinguish among those who have been exposed to tuberculosis but are not infected, those who are chronic carriers, and those who have active disease.

23 Microbial Diseases of the Digestive System

CHAPTER SUMMARY

Structures of the Digestive System (pp. 706–707)

Anatomists commonly divide the structures of the digestive system into two groups: those of the gastrointestinal (GI) tract and those that play an accessory role.

The Gastrointestinal Tract

The GI tract is a long tube lined with mucous membrane and composed of the mouth, esophagus, stomach, small intestine, large intestine, rectum, and anus. The peritoneum surrounds and protects most organs of the GI tract. Food in the tract is moved through the esophagus to the stomach via peristalsis. In the mouth, stomach, and small intestine, food is broken down into nutrients, which are then absorbed, largely via the specialized villi and microvilli lining the small intestine. Absorption of remaining nutrients and water occurs in the large intestine (colon), from which undigested matter, called feces, moves into the rectum for elimination via the anus.

The Accessory Digestive Organs

Accessory digestive organs include the tongue, teeth, liver, gallbladder, and pancreas. The teeth and tongue are important in tearing, mashing, and mixing food into a bolus. The liver produces bile, which is then stored in the gallbladder until required to assist in the breakdown of fats. The liver also breaks down, excretes, or stores toxins; thus, liver damage leads to toxin buildup in the blood. The pancreas produces pancreatic enzymes that assist in digestion, and bicarbonate buffer, which neutralizes stomach acid as it enters the intestine.

Normal Microbiota of the Digestive System (pp. 707)

The esophagus, stomach, and duodenum (the upper portion of the small intestine) are almost sterile. The rest of the GI tract is colonized by a variety of normal microbiota. Bacteria, fungi, and a few protozoa reside on the surfaces and in the pits and crevices of the oral cavity, as well as in saliva. **Viridans streptococci** are the most common normal microbiota of the mouth. The lower small intestine and colon contain an estimated 100 trillion bacteria, which provide some vitamins to the host. They also produce flatus and convert some substances to toxins and cancer-causing chemicals. Their primary benefit is to inhibit pathogens by microbial antagonism.

Bacterial Diseases of the Digestive System (pp. 707–720)

Bacterial digestive diseases are typically mild and self-limiting, but a few can be fatal, especially in infants, the elderly, and the immunocompromised.

Dental Caries, Gingivitis, and Periodontal Disease

Dental caries (cavities) are the second most common infections of humans. Acids produced by mouth bacteria, usually viridans streptococci, eat through the enamel of teeth to create initially painless pits or holes. As the disease progresses, toothache and tooth loss occur. **Gingivitis** is a form of **periodontal disease,** which is inflammation and infection of the tissues surrounding and supporting the teeth. Symptoms include swelling, bleeding, tenderness, and redness of the gums. Tooth loss occurs as the disease progresses. The microbes responsible for dental caries and gingivitis include *Streptococcus mutans,* species of *Lactobacillus,* and *Porphyromonas gingivalis.* Prevention of dental disease requires healthy eating habits (e.g., limiting intake of sugary and sticky foods) and good oral hygiene (brushing and flossing as well as regular dental cleanings). Fluoridation of water is the single most effective way to prevent dental disease within a community.

Peptic Ulcers

Peptic ulcers are erosions of the linings of either the stomach or the duodenum. Abdominal pain, the major symptom, is caused by the invasive action of *Helicobacter pylori.* The microbe burrows through the protective mucus lining the stomach, allowing entry to stomach acid, which can dissolve the lining, creating a hole called a *perforation.* Perforations in turn can lead to internal bleeding and shock requiring immediate medical intervention. Among other virulence factors, *H. pylori* possesses a protein that inhibits acid production by stomach cells, flagella that enable it to burrow through the mucus lining the stomach, and adhesions that facilitate its binding to gastric cells. The route of transmission is thought to be fecal-oral; thus, prevention requires good personal hygiene. Antimicrobial and acid-blocking drugs are used in treatment.

Bacterial Gastroenteritis

Bacterial gastroenteritis (inflammation of the stomach or intestines) is usually caused by consumption of contaminated food or water. Manifestations typically include nausea, vomiting, diarrhea, loss of appetite, abdominal pain, and cramps. In rare cases, kidney failure or anemia can occur. In a severe and painful type of gastroenteritis known as **dysentery,** stools are loose, frequent, and contain mucus and blood. Treatment usually involves replacement of fluids and electrolytes. Antidiarrheal drugs may prolong symptoms by allowing the organisms to remain in the intestines. Good sanitation and personal hygiene are essential preventive measures.

The most common types of bacterial gastroenteritis include the following:

- **Cholera** produces such severe diarrhea that the patient is at risk of dehydration, hypovolemic shock, coma, and death. It is caused by the Gram-negative bacillus *Vibrio cholerae,* which can survive in both saltwater and freshwater supplies. *V. cholerae* produces a potent poison called **cholera toxin,** which stimulates the secretion of excess amounts of electrolytes and water from infected cells. Treatment involves administration of a tetracycline, which reduces the production of the toxin.
- **Shigellosis** is a form of bacterial gastroenteritis in which any of four species of Gram-negative *Shigella* cause diarrhea and the formation of abscesses in the mucosa of the large intestine. Antimicrobial drugs are used in treatment of severe cases, and a vaccine is under development.
- **Traveler's diarrhea** is the common name for bacterial gastroenteritis caused by infection with *Escherichia coli,* a Gram-negative coliform that lives in the

GI tract of humans and other animals. One strain of *E. coli* produces Shiga-like toxin, which is related to the **Shiga toxin** of *Shigella,* and can cause a sometimes fatal gastroenteritis. Shiga-like toxin attaches to the surfaces of neutrophils and is spread by them throughout the body, causing widespread cell death. Antimicrobial drugs induce the production of Shiga-like toxin and thus exacerbate the disease. Treatment is supportive, and no vaccine is available.

- ***Campylobacter* diarrhea** is caused by infection with *Campylobacter jejuni,* a Gram-negative bacterium with polar flagella that possesses adhesions, cytotoxins, and endotoxin (lipid A). *C. jejuni* invades the intestines and produces bleeding lesions and inflammation. Although most cases resolve without treatment, in the United States each year about 100 people die from infection with *C. jejuni.*
- **Antimicrobial-associated diarrhea** is a form of diarrhea that often develops in hospitalized patients taking antimicrobial drugs. The most severe form can result in a life-threatening condition called pseudomembranous colitis caused by the Gram-positive bacterium *Clostridium difficile. C. difficile* produces toxin A and B that trigger intestinal inflammation and lesions that fuse to form a pseudomembrane. Pseudoembranous colitis was rare before the widespread use of antimicrobials and can be prevented by avoiding the unnecessary use of these drugs. Treatment involves the administration of antimicrobial drugs. Fecal transplants and probiotic consumption may help recolonization of the colon with normal microbiota that compete with *C. difficile.*
- **Salmonellosis** and **typhoid fever** are caused by infection with *Salmonella enterica,* a Gram-negative bacillus that lives in the intestines of virtually all vertebrates and is eliminated in their feces. Virulent serotypes of *Salmonella* tolerate the acidic conditions of the stomach, passing into the intestine where they attach via specific adhesins and introduce toxins into host cells. Symptoms include gradually increasing fever, headache, muscle pains, and malaise; a rash may appear. With typhoid fever, life-threatening complications such as intestinal hemorrhage and *peritonitis* (inflammation of the peritoneum) may occur. Humans acquire the infection via consumption of contaminated food or water. Salmonellosis is generally self-limiting, whereas 12–30% of typhoid fever patients die without administration of antimicrobial drugs. Prevention centers on good hygiene, especially in the kitchen.

Bacterial Food Poisoning (Intoxications)

Food poisoning is a broad term used to refer to either consuming pathogens (gastroenteritis) or their toxins (*intoxications*). General symptoms of bacterial intoxication include nausea, vomiting, diarrhea, abdominal cramping, discomfort, bloating, loss of appetite, and fever. Most cases are self-limiting and resolve within 24 hours. A common culprit is *Staphylococcus aureus,* which may be introduced into a food during preparation. Its five enterotoxins induce the characteristic symptoms. Because the toxins are heat-stable, warming or reheating a food does not destroy them. Treatment is administration of electrolyte-balanced fluids to prevent dehydration. Good kitchen hygiene is preventive.

Viral Diseases of the Digestive System (pp. 720–726)

Viral diseases of the digestive system include oral herpes, mumps, viral gastroenteritis, and viral hepatitis.

Oral Herpes

Oral herpes is among the most common of the infections caused by the *Herpesviridae* family. Its symptoms include painful, itchy, creeping lesions, called **fever blisters** or **cold sores,** on the lips, accompanied by fever, malaise, and muscle pain. Infection is lifelong and outbreaks recur, but the lesions are milder. Severe infection extending into the oral cavity is called *herpetic gingivostomatitis,* and infection affecting the esophagus is *herpes esophagitis.* HHV-1 accounts for 90% of cases, and is transmitted typically via casual contact during childhood. Topical creams containing acyclovir limit symptoms, but there is no cure. Prevention requires avoiding contact with infected individuals.

Mumps

Effective immunization has made **mumps** nearly extinct in developed nations, but epidemics still occur in nations with inadequate immunization programs. Infection with a genus of *Rubulavirus,* commonly known as the mumps virus, causes *parotitis* (swelling of the parotid salivary glands), face pain, fever, headache, and sore throat. There is no treatment, but recovered individuals have lifelong immunity.

Viral Gastroenteritis

Although the general manifestations of **viral gastroenteritis** are the same as those of bacterial gastroenteritis, viral forms are usually less severe than bacterial forms of the disease. Symptoms usually resolve within 12–60 hours. Dehydration is the most common complication. Three viruses—*caliciviruses* such as **noroviruses,** *astroviruses,* and *rotaviruses*—are commonly causative. These enter the digestive tract via contaminated food or water and infect the cells lining the intestinal tract, where they undergo lytic replication. There is no specific treatment, other than replacement of fluids and electrolytes to prevent dehydration. Oral vaccines against rotaviruses are available.

Viral Hepatitis

Hepatitis is inflammation of the liver produced by autoimmune disease, alcohol or drug abuse, genetic disorders, or microbial infection. Symptoms include jaundice, dark urine, abdominal pain and distention, nausea, vomiting, weight loss, fatigue, and fever. Liver failure can occur with chronic infection. Immune responses that kill infected host cells cause most of the damage to the liver of infected patients.

Five viruses responsible for most cases of viral hepatitis include: Hepatovirus Hepatitis A virus (HAV), Orthohepadnavirus Hepatitis B virus (HBV), Hepacivirus Hepatitis C virus (HCV), Deltavirus Hepatitis delta virus (HDV), and Hepevirus Hepatitis E virus (HEV). Hepatitis A and E, which are spread via the fecal-oral route, are usually cleared via the immune response. The other viruses, which are spread sexually and via contaminated needles, typically remain, resulting in chronic infection. An association between HBV infection and hepatic (liver) cancer is based on strong medical evidence. Treatment involves rest and reducing inflammation; there is no cure. Immunoglobulin given immediately after exposure offers some protection. Hygienic measures are essential to prevent transmission. Vaccines are available against hepatitis A and B.

Protozoan Diseases of the Intestinal Tract (pp. 727–731)

Relatively few protozoa cause infection of the GI tract.

Giardiasis

Giardiasis is one of the more common water-borne GI diseases in the United States. Symptoms include one to two weeks of watery diarrhea, abdominal pain, flatus, nausea and vomiting, and a low-grade fever. Patients' stools smell of hydrogen sulfide. A diplomonad flagellate named *Giardia intestinalis,* a common resident of waterways throughout the United States, causes giardiasis. Upon consumption of contaminated water, ingested cysts release trophozoites in the small intestine. These multiply and interfere with intestinal absorption of nutrients. Treatment is with antimicrobial drugs and supportive care to replace lost fluids. Neither humans nor their pets should drink unfiltered stream or river water, and individuals recovering from giardiasis must be extremely vigilant about their personal hygiene.

Cryptosporidiosis

Infection with the protozoan *Cryptosporidium parvum* causes **cryptosporidiosis,** a disease characterized by severe watery diarrhea several times a day lasting about two weeks. Severe fluid and weight loss can accompany the diarrhea, and life-threatening malabsorption, hepatitis, and pancreatitis can complicate the disease. Infection most commonly results from drinking water contaminated with oocysts, but direct fecal-oral transmission also occurs. Fluid and electrolyte replacement is the primary treatment. Prevention requires good hygiene, avoiding consumption of contaminated food or water, and avoiding fecal exposure during sex.

Amebiasis

Amoebae are protozoa with no defined shape that acquire food with pseudopodia. Infection with *Entamoeba histolytica,* a motile trophozoite, can result in one of three forms of **amebiasis:** *Luminal amebiasis* is generally asymptomatic; *amebic dysentery* is characterized by severe diarrhea, colitis, appendicitis, ulceration of the intestinal mucosa, bloody stools, and/or pain; and *invasive extraintestinal amebiasis* causes potentially fatal lesions of dead and dying intestinal cells in the liver, lungs, spleen, kidneys, or brain.

Infection typically begins with consumption of contaminated food or water, or ingestion from contaminated hands or during oral-anal intercourse. Treatment involves oral rehydration therapy and antiamebic drugs. Preventive measures include avoiding uncooked foods and drinking bottled water in endemic areas, as well as effective water processing and good personal hygiene.

Helminthic Infestations of the Intestinal Tract (pp. 731–734)

Helminths can infest the GI tract as non–disease-causing parasites.

Tapeworm Infestations

Tapeworm is the common name for a **cestode,** which is a flat, segmented, parasitic helminth. Tapeworm infestations are usually asymptomatic, and a person does not recognize they carry a worm unless passing segments of the helminth. Rarely, infestation is accompanied by nausea, abdominal pain, weight loss, and diarrhea, and long worms occasionally block the intestines, preventing normal bowel function. The common tapeworms of humans are *Taenia saginata,* the beef tapeworm, and *Taenia solium,* the pork tapeworm. Tapeworms attach to host tissue via the **scolex,** a small organ that possesses suckers and/or hooks. Behind the scolex is the neck region from which a chain (called a *strobila*) of body segments called **proglottids** grows continuously.

Monoecious proglottids are fertilized to become gravid with eggs and shed into the environment, where they are consumed by intermediate hosts. Eggs hatch in these animals and eventually form **cysticerci** (cysts) in muscle tissue. Humans are infected by consuming cysticerci in meat. Treatment is with niclosamide or praziquantel. Rarely, surgery is necessary to remove tapeworms from the intestinal tract. Thoroughly cooking or freezing meat is the easiest method of prevention.

Pinworm Infestations

Infestation with *Enterobius vermicularis*—commonly known as **pinworm**—causes intense perianal itching, irritability and sleep disturbance due to itching, decreased appetite, and possibly weight loss. *Enterobius* is a **nematode,** a long, thin, unsegmented cylindrical helminth tapering to points at each end. Male and female worms mate in the intestinal tract, and the female crawls out of the anus to lay eggs. Treatment is with pyrantel pamoate or mebendazole. Strict personal hygiene prevents reinfection.

QUESTIONS FOR FURTHER REVIEW

Answers to these questions can be found in the answer section at the back of this study guide. Refer to the answers only after you have attempted to solve the questions on your own.

Multiple Choice

1. Which of the following is not an accessory digestive organ?
   a. pancreas
   b. tongue
   c. gallbladder
   d. esophagus
2. The majority of food digestion occurs in the
   a. stomach.
   b. small intestine.
   c. large intestine.
   d. None of the above is correct.
3. Where in the body is bile stored?
   a. liver
   b. pancreas
   c. gallbladder
   d. stomach
4. Which of the following regions of the digestive system is colonized by microorganisms?
   a. colon
   b. stomach
   c. esophagus
   d. duodenum
5. Which condition falls behind only the common cold in its frequency of occurrence?
   a. gingivitis
   b. dental caries
   c. gastroenteritis
   d. periodontitis
6. An ulcer that pierces the stomach is referred to as a
   a. uodenal ulcer.
   b. gastric ulcer.
   c. perforation.
   d. None of the above is correct.
7. Which of the following is a standard treatment for gastroenteritis?
   a. antidiarrheal drugs
   b. fluid and electrolyte replacement
   c. antimicrobial drugs
   d. All the above are correct.

8. Cholera is often diagnosed by the presence of
   a. bloody stool.
   b. purulent stool.
   c. abdominal pain.
   d. rice-water stool.

9. The most common pathogen causing diarrhea is
   a. *Escherichia coli.*
   b. *Vibrio cholera.*
   c. *Campylobacter jejuni.*
   d. *Shigella dysenteriae.*

10. A vaccine exists for which species of *Shigella*?
    a. *Shigella dysenteriae*
    b. *Shigella boydii*
    c. *Shigella sonnei*
    d. None of the above is correct.

11. What drug would likely be prescribed to someone with oral herpes?
    a. Acyclovir
    b. Metronidazole
    c. Paromomycin
    d. Azithromycin

12. Viral gastroenteritis is most frequent in what season?
    a. summer
    b. winter
    c. fall
    d. spring

13. Hepatitis delta virus can only spread from cells in the presence of
    a. Hepatitis C virus.
    b. Hepatitis A virus.
    c. Hepatitis B virus.
    d. Hepatitis E virus.

14. Complete infectious virions of hepatitis B virus are called
    a. Dane particles.
    b. filamentous particles.
    c. spherical particles.
    d. none of the above.

15. For which of the following viruses is a vaccine not available?
    a. Hepatitis A virus
    b. Hepatitis C virus
    c. Hepatitis B virus
    d. All the above are correct.

16. The part of the tapeworm that is involved in attachment is the
    a. proglottids.
    b. cuticle.
    c. scolex.
    d. None of the above is correct.

17. The type of amebiasis that is asymptomatic is
    a. invasive extraintestinal amebiasis.
    b. amebic dysentery.
    c. noninvasive extraintestinal amebiasis.
    d. luminal amebiasis.

18. Tapeworm infestation is typically treated with
    a. niclosamide.
    b. mebendazole.
    c. pyrantel pamoate.
    d. iodoquinol.

Fill in the Blanks

1. The membrane covering most of the gastrointestinal tract organs is called the ____________________.

2. ____________________ is the term for the muscle contractions that move food down the esophagus.

3. The most prevalent microbes of the mouth are ____________________.

4. Pathogens in the body that provide a protective effect by outcompeting pathogens are referred to as ____________________.

5. The most effective method used to prevent dental diseases is ____________________.

6. *Helicobacter pylori* neutralizes stomach acid by secreting ____________________.

7. Inflammation of the stomach or intestines due to bacteria is called ____________________.

8. ____________________ are toxins that bind to proteins lining the surface of the intestines.

9. The skin lesion on the lips of an individual with oral herpes is commonly called a(n) ____________________.

10. HHV-1 becomes latent in cells known as ____________________.

11. Inflammation of the liver is referred to as ____________________.

12. ____________________ is the most common route of transmission of hepatitis B virus (HBV).

13. The most common form of amebiasis is ____________________.

14. A cestode is more commonly called a(n) ____________________.

Matching

Match the organism on the left to its corresponding disease on the right. Each letter will be used only once.

1. ____ *Salmonella enterica* Enteritidis
2. ____ *Porphyromonas gingivalis*
3. ____ Norovirus
4. ____ *Streptococcus mutans*
5. ____ Herpes simplex virus type 1
6. ____ *Salmonella enterica* Typhi
7. ____ *Helicobacter pylori*
8. ____ *Escherichia coli*
9. ____ *Enterobius vermicularis*
10. ____ *Staphylococcus aureus*

A. Traveler's diarrhea
B. Dental caries
C. Typhoid fever
D. Peptic ulcer
E. Bacterial food poisoning
F. Salmonellosis
G. Viral gastroenteritis
H. Periodontitis
I. Pinworm infestation
J. Cold sores

Short-Answer Questions for Thought and Review

1. Summarize the mechanisms of action of cholera toxin. In *Vibrio cholerae* infections, what is the main cause of death?

2. How does HBV protect itself from immune clearance?

3. Bacterial food poisoning due to intoxication can proceed rapidly, with symptoms resolved within 24 hours. Considering the nature of the illness, give a possible explanation for the rapid onset and disappearance of symptoms.

Critical Thinking

1. Antimicrobial agents are generally not prescribed for enteric diarrheal diseases because such infections are often self-limiting. However, other reasons exist for not prescribing antimicrobials in this situation. Suggest some of these reasons.
2. You are out hiking in the mountains and come upon a clear mountain stream. You're out of water and don't have a filtration kit with you. There is a beaver dam across the stream but you don't see any beavers or other signs of beaver activity. Is it safe to drink the water from the stream if you do so upstream of the dam?
3. Two children are planning to visit the dentist. One lives deep in the countryside and her family is responsible for their water supply. The other child lives in an urban setting with water from a municipal source. Which child would you predict to have more cavities and why?

Concept Building Questions

1. Why are cysts more likely to be the infective form rather than trophozoites when dealing with protozoal parasites of the digestive system? Consider structure and mode of transmission in your answer.
2. Provide reasons why bacterial gastroenteritis is associated more often with developing countries than developed countries.

24 Microbial Diseases of the Urinary and Reproductive Systems

CHAPTER SUMMARY

Structures of the Urinary and Reproductive Systems (pp. 740–742)

The urinary and reproductive systems in females are anatomically distinct. In males, the two systems share some structures.

Structures of the Urinary System

In both males and females, the largest and most superior organs of the urinary system are the two kidneys, which are composed of millions of *nephrons* that filter blood to form urine. Each kidney has about 1.25 million nephrons, each of which is composed of a ball of capillaries called a *glomerulus*. The kidneys are connected by ureters to the urinary bladder, which stores urine until it is excreted via the urethra.

Structures of the Reproductive System

The interior organs of the female reproductive system include two ovaries, which produce haploid ova, one of which is typically released monthly. Ova are swept through either of two uterine tubes toward the uterus (womb), the lining of which is thickened after ovulation in preparation for pregnancy. The uterus ends inferiorly at the cervix, which is in turn superior to the mucous-membrane-lined vagina (birth canal). The external genitalia include the clitoris, two sets of labia (lips), and the opening of the vagina.

Male reproductive organs include two testes (testicles) located in an external pouch called the scrotum, a system of ducts, accessory glands, and the penis. Testes produce sperm cells that are stored in the epididymis and pass through the ductus deferens and into the urethra, which passes through the prostate gland and into the penis.

If a sperm cell fertilizes an ovum, the resulting diploid zygote develops into an embryo that implants in the uterine wall and further develops into a fetus who passes through the cervix and the birth canal to be born approximately nine months from fertilization. When an ovum is not fertilized, it passes with the lining of the uterus out of the vagina during menstruation.

In females, the vaginal opening can be a portal for microorganisms, especially during sexual intercourse. In males, microbes can enter the body through the urethra or skin of the penis.

Normal Microbiota of the Urinary and Reproductive Systems

The urethra normally supports the growth of some microbiota, chiefly avirulent species of *Lactobacillus, Staphylococcus,* and *Streptococcus.* In both males and females, the rest of the urinary organs and the urine in them are sterile due to the normally acidic pH of urine and the flushing action of urination. The vagina is home to a wide variety of normal microbiota that vary with estrogen levels.

Bacterial Diseases of the Urinary System (pp. 742–744)

Both **urinary tract infections** (UTIs) and systemic bacterial diseases can affect the health of the urinary system.

Bacterial Urinary Tract Infections

Invading bacteria may trigger inflammation in the urethra (**urethritis**), urinary bladder (**cystitis**), prostate (**prostatitis**), or kidneys (**pyelonephritis**). Mild UTIs may produce no symptoms or only a slight fever, but most often UTIs involve frequent, urgent, and painful urination, called *dysuria*. Pyelonephritis may trigger severe abdominal, flank (side), and back pain, high fever, chills, vomiting, and fatigue. If untreated, bacteria may enter the blood (*bacteremia*) and prove fatal. Gram-negative enteric bacteria such as *Escherichia coli* are commonly the causative pathogens. UTIs are more common in females than males because their urethra is much shorter and is closer to the anus, promoting introduction of fecal bacteria into the urinary system. No treatment is necessary for mild cases, but antimicrobial drugs prevent the spread of infection to the kidneys and blood. Intravenous administration of antimicrobials is used for severe infection. Prevention in females includes wiping from front to back after defecation, urinating following sexual intercourse, refraining from douching, and using a form of birth control other than diaphragms. Drinking 2–4 liters of fluid per day is also encouraged.

Leptospirosis

Leptospirosis is a **zoonosis;** that is, a disease primarily seen in animals that spreads to humans. The causative agent enters the body through breaks in the skin or mucous membranes and spreads to the urinary system from the blood. An abrupt fever follows, accompanied by myalgia (muscle pain), muscle stiffness, and headache. Half of patients develop nausea, vomiting, and diarrhea. Kidney and liver failure, meningitis, and respiratory distress are rare but potentially fatal complications. Taxonomists currently believe that all strains of the Gram-negative spirochete causing leptospirosis belong to a single species: *Leptospira interrogans*. Treatment is with oral antibiotics or intravenous penicillin for severe infections, and a vaccine is available for livestock and pets.

Streptococcal Acute Glomerulonephritis

When some strains of group A *Streptococcus* infect adults, antibody-antigen complexes can accumulate in the glomeruli of the kidneys to trigger streptococcal acute **glomerulonephritis,** a progressive and irreversible kidney disease. Young patients usually recover fully.

Nonvenereal Diseases of the Reproductive System (pp. 744–747)

Nonvenereal diseases of the reproductive system include urethritis and prostatitis in males (already discussed) as well as staphylococcal toxic shock syndrome, bacterial vaginosis, and candidiasis in females.

Staphylococcal Toxic Shock Syndrome

Sudden onset of fever, chills, vomiting, diarrhea, extremely low blood pressure, mental confusion, and a severe rash characterize **staphylococcal toxic shock syndrome** (STSS). Untreated STSS is frequently fatal. Certain strains of *Staphylococcus aureus* cause the condition, most notably in menstruating females who use super-absorbent tampons, by producing exotoxins. STSS is a medical emergency; treatment is with antimicrobial drugs and immunoglobulin to counteract the toxin. Women can reduce their risk by avoiding the use of tampons or by choosing less absorbent tampons and changing them frequently. Vaginal sponges and diaphragms for birth control are also to be avoided or used intermittently for short periods.

Bacterial Vaginosis

Bacteria infecting the warm, moist vaginal lining cause **bacterial vaginosis.** The condition is characterized by a white discharge with a "fishy" odor. Some itching or irritation may also occur. Up to half of cases are asymptomatic. Infection results when the normal lactobacilli of the vagina are replaced with a large number of facultatively or obligate anaerobic bacteria. A decline in the number of lactobacilli results in a pH higher than the normal 4.5, which either promotes or allows the growth of the pathogenic bacteria. Because douching reduces the normal population of lactobacilli in the vagina, it is a primary risk factor. A history of multiple sexual partners is also a risk factor. Treatment is with antibacterial drugs. Other than avoidance of douching and abstinence, there are no known preventive measures.

Candidiasis

Candidiasis describes any of a variety of opportunistic yeast infections and diseases caused by various species of the genus *Candida*. Vaginal candidiasis manifests as white mucoid colonies growing on the mucous membrane of the vagina and the skin covering the labia. The yeast produces burning and itching, and intercourse can be painful. The discharge is often curdlike and slight. *Candida albicans* is the most common species causing candidiasis. It forms long cellular extensions called **pseudohyphae.** Treatment is with antifungal agents. Because antibacterial agents suppress microbial antagonism, allowing *C. albicans* to flourish, unnecessary use of antibacterial drugs is discouraged.

Sexually Transmitted Infections (STIs) and Diseases (STDs) (pp. 747–749)

The World Health Organization estimates that 333 million new cases of STDs occur each year. About a third of all STDs affect people under age 25. Female adolescents are at significant risk because their cervical lining is especially prone to bacterial invasion, and yet STDs in young women are more likely to be asymptomatic. Consequences of STDs common in this population include **pelvic inflammatory disease** (PID)—inflammation in the uterus, uterine tubes, or ovaries—which can result in ectopic pregnancies, sterility, cervical cancer, and other health problems.

Bacterial STDs (pp. 749–757)

Bacterial infections are among the more familiar STDs.

Gonorrhea

Infection with the Gram-negative bacterium *Neisseria gonorrhoeae* causes **gonorrhea,** a disease recognized for centuries. In men, gonorrhea causes acute inflammation, extremely painful urination, and a purulent discharge. In women, the infection is often asymptomatic while silently damaging the uterine tubes and resulting in sterility. It can also cause PID. Newborns infected during vaginal birth may suffer inflammation of the cornea, potentially leading to blindness.

Gonococci adhere via their fimbriae and capsules to epithelial cells of the mucous membranes lining much of the genital, urinary, and digestive tracts of humans. The microbe can also attach to sperm cells, thereby invading the sexual partner's uterus, uterine tubes, and beyond. Phagocytized bacteria survive and multiply within neutrophils. Treatment is with broad-spectrum oral cephalosporin. No vaccine is available. Efforts to stem the spread of gonorrhea focus on education to change sexual behavior, aggressive detection, and the screening of all sexual contacts of carriers.

Syphilis

Infection with *Treponema pallidum,* a spirochete, causes **syphilis.** *T. pallidum* is an obligate parasite of humans and is transmitted almost solely by sexual contact. It can also be transmitted from a pregnant woman to her fetus, and rarely through blood transfusion. Syphilis has four phases:

- In **primary syphilis,** a small, hard, extremely infectious **chancre** (lesion) fills with spirochetes at the site of infection and remains for several weeks. In about a third of cases, disappearance of the chancre is the end of the disease.
- In about two-thirds of cases, *Treponema* invades the bloodstream and causes the symptoms of **secondary syphilis,** including malaise and a long-lasting contagious skin rash. Rash lesions are filled with contagious spirochetes.
- After several weeks or months, the rash disappears and the patient enters an asymptomatic stage called **latent syphilis.** The majority of cases do not advance further.
- About a third of patients advance to **tertiary syphilis,** a hyperimmune response characterized by swollen lesions called **gummas** on the skin or other organs, destruction of cardiovascular or central nervous system tissue, personality changes, insanity, or blindness.

When *Treponema* crosses the placenta from an infected mother to her fetus, **congenital syphilis** results. Spontaneous abortion may occur, or the infant may be born with mental retardation or malformed organs. Treatment of all forms except tertiary syphilis is with penicillin G. No vaccine is available.

Chlamydial Infections

The most common sexually transmitted bacterium is *Chlamydia trachomatis.* In males, infection can cause urethritis, painful urination, purulent discharge, **epididymitis** (inflammation of the epididymis) or **orchitis** (inflammation of a testis), and sterility. Infection is often asymptomatic in females. When newborns are infected at birth, an eye disease called **trachoma** develops. Trachoma is the leading nontraumatic cause of blindness worldwide. **Lymphogranuloma venereum** is a severe form of chlamydial STD characterized by a transient genital lesion at the site of infection, followed by development of a **bubo** (a painfully enlarged lymph node) in the groin, accompanied by fever, chills, anorexia, and muscle pain. Antimicrobial drugs are administered for genital infections, and newborns with trachoma are treated with erythromycin cream. Surgical correction of eyelid deformities may be necessary. Abstinence or mutual monogamy are the only preventives.

Chancroid

Chancroid is characterized by soft, painful genital ulcers called soft chancres that form at the site of infection. The condition may be asymptomatic; however, the most common symptom in women is painful urination caused by the presence of a chancroid ulcer blocking the opening of the urethra. In half of patients, the infection spreads to the lymph nodes, producing buboes. Chancroid is caused by *Haemophilus ducreyi,* a Gram-negative bacterium that is an obligate parasite of humans and produces a toxin that kills epithelial cells. Treatment is with antimicrobial drugs. Buboes need to be drained. As with all STDs, abstinence or mutual monogamy are the best preventives.

Viral STDs (pp. 757–760)

Viruses are the most common causes of STDs, including AIDS, which is discussed in Chapter 18.

Genital Herpes

Genital herpes manifests as numerous small blisters on the genitals, around the rectum, or on adjacent areas of skin. The blisters burn and itch before breaking to become painful ulcers. Patients experience fever, myalgia, malaise, and decreased appetite. Two enveloped dsDNA viruses called human herpes viruses (HHV) 1 and 2 cause herpes. About 85% of cases involve HHV-2. HHV-1 (oral herpesvirus) causes the remaining cases, usually following oral-genital contact. Infected people can shed herpesviruses in mucous secretions even in the absence of lesions. The virus typically enters a latent phase in nerve ganglia, becoming active and triggering recurrent, though less severe, episodes for years. Herpesviruses also spread by nonsexual contact, such as during vaginal birth, or via contact with herpes blisters on the skin, a condition called *whitlow.* There is no cure; however, symptoms can be relieved with administration of acyclovir. Abstinence or monogamy is preventive. Cesarean birth is required if a pregnant woman has lesions at the time of delivery.

Genital Warts

Infection with *human papillomaviruses (HPVs)* causes **genital warts,** which range in size from barely visible to cauliflower-like growths called **condylomata acuminata.** Some HPVs may cause cancer. Genital warts may be itchy or painful, and rarely they may bleed or cause increased vaginal discharge. Transmission is via the skin or mucous membranes of the penis, vagina, or anus during sexual intercourse. Physicians remove genital warts with surgery, freezing, burning, laser, or the use of caustic chemicals, though with any of these methods, viruses may remain in surrounding tissues and warts may return. In women, the Papanicolaou smear (Pap smear) screens for cervical cancer. Abstinence or monogamy is preventive. A vaccine is available.

Protozoan STDs (pp. 760–761)

Sexual activity can transmit several protozoa, including intestinal parasites such as *Giardia* and *Cryptosporidium;* however, this is not the regular way these protozoa are infective. The protozoan *Trichomonas* normally infects the reproductive system.

Trichomoniasis

Trichomonads are flagellated protozoan parasites of animals and humans. *Trichomonas vaginalis* is the one pathogenic species, causing an STD called **trichomoniasis.** Infection is asymptomatic in males. In females, it causes a foul-smelling, yellow-green vaginal discharge and vaginal irritation. There may also be lesions of the genitalia, abdominal pain, dysuria, and pain during sexual intercourse. Rarely, infected males develop urethritis or prostatitis. The fetus of an infected mother can become infected during vaginal birth. Treatment involves the patient and all sexual contacts, and is with metronidazole taken orally. Babies are protected by treating infected mothers with metronidazole before they give birth. Prevention requires abstinence or monogamy.

QUESTIONS FOR FURTHER REVIEW

Answers to these questions can be found in the answer section at the back of this study guide. Refer to the answers only after you have attempted to solve the questions on your own.

Multiple Choice

1. Which of the following is not a component of the male urinary tract?
   a. urethra
   b. prostate gland
   c. kidneys
   d. These are all part of the male urinary tract.

2. The functional unit of the kidney that filters the blood is called
   a. nephrons.
   b. renal cortex.
   c. renal pyramids.
   d. none of the above.

3. Where are eggs produced in the female reproductive system?
   a. uterine tubes
   b. uterus
   c. vagina
   d. ovaries

4. Where is sperm produced in the male reproductive system?
   a. ductus deferens
   b. testes
   c. prostate gland
   d. scrotum

5. Which of the following sites would usually be sterile?
   a. urethra
   b. vagina
   c. urinary bladder
   d. All of the above are correct.

6. Inflammation of the urinary bladder is called
   a. cystitis.
   b. urethritis.
   c. pyelonephritis.
   d. prostatitis.

7. Which of the following is a zoonotic disease?
   a. candidiasis
   b. chancroid
   c. leptospirosis
   d. trichomoniasis

8. Which of the following is a nonvenereal disease of the reproductive system?
   a. genital warts
   b. trichomoniasis
   c. leptospirosis
   d. candidiasis

9. Women who use the following are at risk for toxic shock syndrome.
   a. highly absorbent tampons
   b. vaginal sponges
   c. diaphragms
   d. All the above are correct.

10. Bacteria that infect the lining of the vagina cause a condition known as
    a. candidiasis.
    b. vaginosis.
    c. vaginitis.
    d. none of the above.

11. Which of the following is not a complication seen with untreated pelvic inflammatory disease?
    a. ectopic pregnancy
    b. sterility
    c. cervical cancer
    d. These are all possible complications of PID.

12. Gonorrheal infection in women usually produces
    a. asymptomatic infection.
    b. painful urination.
    c. purulent discharge.
    d. none of the above.

13. A chancre forms during
    a. latent syphilis.
    b. secondary syphilis.
    c. primary syphilis.
    d. tertiary syphilis.

14. This microbe has not been cultured in cell-free media.
    a. *Neisseria gonorrhoeae*
    b. *Treponema pallidum*
    c. *Leptospira interrogans*
    d. *Chlamydia trachomatis*

15. The most common reportable sexually transmitted disease in the United States is
    a. gonorrhea.
    b. syphilis.
    c. chancroid.
    d. chlamydia.

16. *Neisseria gonorrhoeae* cells that lack which of the following are avirulent?
    a. lipooligosaccharide
    b. fimbriae
    c. polysaccharide capsule
    d. All the above are correct.

17. The most common cause of sexually transmitted diseases is
    a. viruses.
    b. fungi.
    c. bacteria.
    d. protozoa.

18. The leading cause of nontraumatic blindness is caused by this sexually transmitted pathogen.
    a. *Neisseria gonorrhoeae*
    b. *Treponema pallidum*
    c. *Chlamydia trachomatis*
    d. *Leptospira interrogans*

Fill in the Blanks

1. An embryo implants into the wall of the ____________________.
2. ____________________ is the condition characterized by frequent, urgent, and painful urination.
3. The most common cause of urinary tract infections is intestinal microbiota referred to as ____________________.
4. Streptococcal group A antibody-antigen complexes that deposit in the kidneys can cause ____________________.

5. Toxic shock syndrome toxins bind to ____________________ and ____________________ in the immune system of the host, triggering the signs and symptoms of toxic shock syndrome.

6. *Candida albicans* forms long cellular extensions called ____________________.

7. In women, the part of the reproductive system that *N. gonorrhoeae* most commonly infects is the ____________________.

8. Gummas are characteristic of ____________________.

9. Most cases of syphilis do not progress beyond the stage called ____________________.

10. Bacteria of the genus ____________________ do not have cell walls.

11. The lesions characteristic of chancroid are referred to as ____________________.

12. ____________________ are the giant growths that sometimes occur when an individual has genital warts.

13. The most common curable STD of women is ____________________.

Matching

Match the organism on the left to its corresponding disease on the right. Each letter will be used only once.

1. ___ *Staphylococcus aureus*
2. ___ *Mycoplasma hominis*
3. ___ Papillomavirus
4. ___ *Treponema pallidum*
5. ___ Herpes simplex virus type 2
6. ___ *Chlamydia trachomatis*
7. ___ *Haemophilus ducreyi*
8. ___ *Escherichia coli*

A. Genital warts
B. Lymphogranuloma venereum
C. Bacterial urinary tract infection
D. Chancroid
E. Genital herpes
F. Toxic shock syndrome
G. Syphilis
H. Bacterial vaginosis

Short-Answer Questions for Thought and Review

1. Urinary tract infections are more common in females than males. Give some reasons for this phenomenon.

2. Is condom use a good measure to prevent women from transmitting genital herpes to a partner? Why or why not?

3. Briefly explain how toxic shock syndrome toxins (TSSTs) lead to the signs and symptoms of toxic shock syndrome.

Critical Thinking

1. An individual went to a health clinic to be tested for syphilis. A sample was taken and sent to a lab to be grown on an agar plate to determine the presence of the causative agent. Nothing grew on the agar plate; would you diagnose this person as not having syphilis? Explain your answer.
2. To prevent neonatal herpes, babies are delivered by cesarean section if the mother has active genital lesions at the time of birth. If she does not have active lesions but is still infected, why isn't it absolutely necessary to do a cesarean section at birth?
3. Explain the life cycle of *Chlamydia trachomatis*. What is unique about the life cycle of this bacterium?

Concept Building Questions

1. Many sexually transmitted diseases, like *Chlamydia* and *Treponema* infections, are either asymptomatic or produce symptoms that are not always obvious. How is this beneficial to the bacteria, but disadvantageous to health care workers?
2. An individual with tertiary syphilis is prescribed antibiotics to try to eliminate symptoms. Will this course of treatment work?

25 Applied and Environmental Microbiology

CHAPTER SUMMARY

Applied microbiology is the commercial use of microorganisms. It is divided into following two types: **food microbiology** and **industrial microbiology**. **Environmental microbiology** involves the study of microorganisms found in nature.

Food Microbiology (pp. 768–775)

Food microbiology encompasses the use of microorganisms to produce food and to control microbial activity that results in food spoilage.

The Roles of Microorganisms in Food Production

A classic example of microorganisms in food production is leavened bread, in which the aerobic metabolism of the yeast *Saccharomyces cerevisiae* causes dough to rise. *Fermentation* refers to desirable changes that occur to a food or beverage as a result of microbial growth. In contrast, **spoilage** involves unwanted changes to a food due to the action of microorganisms. Sourdough breads use **starter cultures** made of yeast and lactic acid bacteria to give a characteristic flavor; such cultures are composed of known microorganisms that perform specific fermentations consistently. *Secondary cultures* may be added to further modify the flavor or aroma of foods. Fermentation is used in the production of many food products, such as *sauerkraut*, *kimchi*, and *soy sauce*, as well as chocolate and coffee. *Pickling* is the process of preserving or flavoring foods with brine or acid, which can come from microbial fermentation. *Silage* is an animal feed made from fermented grains and other foliage. Fermentation is also used, sometimes in combination with smoking or drying, to preserve meats and fish. Many dairy products—including buttermilk, yogurt, and cheese—utilize starter cultures of lactic acid bacteria to give foods their characteristic textures and flavors.

Alcoholic fermentation is the process by which various species of microorganisms convert simple sugars into alcohol and carbon dioxide. Starter cultures of yeast are often used in the fermentation process. *Saccharomyces cerevisiae* ferments sugar in fruit juice to alcohol for winemaking. Distilled spirits are made similarly to wines but concentrate the alcohol by distillation. Beer is made from barley and utilizes either bottom-fermenting yeast or top-fermenting yeast for fermentation. Sake is made similarly to beer but uses rice as a base. Vinegar is produced when the ethanol from fermentation is oxidized to acetic acid by the action of *acetic acid bacteria*.

The Causes and Prevention of Food Spoilage

Food spoilage may be due to *intrinsic factors* or *extrinsic factors*. Intrinsic factors include the nutritional composition of foods, water activity, acidity, physical structure, and microbial competition. Some foods contain natural antimicrobial agents, whereas others, such as *fortified foods*, may facilitate the growth of microorganisms by providing more nutrients. Moist foods are more susceptible to spoilage than dry foods, and low pH supports little microbial growth. Rinds or skins protect some fruits and vegetables; damaging these coverings or grinding foods (such as meats) allows access to the moist interior and provides more surfaces for microbes to grow. Microbial competition in fermented foods also retards spoilage. Extrinsic factors include the ways in which food is processed, handled, and stored.

Foods may be *perishable*, *semiperishable*, or *nonperishable*. *Industrial canning* is a major food packaging method for preserving foods. The two most frequent contaminants in canned foods are *Clostridium* species and coliforms. *Pasteurization* is used with beer, wine, and dairy products and is less rigorous than canning; pasteurized foods tend to spoil without refrigeration. Drying or dessication utilizes ovens or heated drums to evaporate water from foods and to reduce microbial growth. *Lyophilization*, or *freeze-drying*, involves freezing foods and then using a vacuum to draw off ice crystals. *Irradiation* with gamma radiation, generally from the isotope cobalt-60, can achieve complete sterilization of foods but is controversial. UV light is often used to treat packing and cooling water and surfaces used in food processing. *Aseptic packaging* involves sterilizing packaging materials and sealing foods inside to preserve freshness.

Salt and sugar can kill microbes present during food processing and retard further microbial growth. Some natural preservatives include allicin in garlic and benzoic acid in cranberries. Wood smoke introduces growth inhibitors that help preserve some foods. Other chemicals are purposely added to foods as preservatives, including organic acids and gases. These are typically *germistatic*—that is, they inhibit microbes' growth—rather than *germicidal* (killing microbes). Higher temperatures are generally best for food processing and lower temperatures for food storage. Cold temperatures retard growth; freezing does not kill all microorganisms but may lower microbial contamination to reduce the chance of food poisoning. *Listeria monocytogenes*, which causes listeriosis, is one microbe that continues to grow under refrigeration.

Foodborne Illnesses

Consuming spoiled food can cause illness, but not all foodborne illnesses result from food spoilage. *Food poisoning* can be divided into two types: *food infections* are caused by the consumption of living microorganisms, whereas *food intoxications* are caused by the consumption of microbial toxins. Signs and symptoms for both include nausea, vomiting, diarrhea, fever, fatigue, and muscle cramp 2–48 hours after ingestion. Most outbreaks are *common-source epidemics*, with one food source responsible. More than 250 different foodborne diseases have been described. Most are caused by bacteria.

Industrial Microbiology (pp. 775–785)

The Roles of Microbes in Industrial Fermentations

Industrial fermentations involve the large-scale growth of particular microbes for producing beneficial compounds, such as amino acids or vitamins. The process is performed in huge, sterilizable vats, often using waste products from other industrial or food processes. In *batch production*, organisms ferment their substrate until it is exhausted, and then the end product is harvested; *continuous flow production* involves the continuous addition of new medium. Industrial products are either *primary metabolites*, which are produced during active growth and metabolism, or *secondary metabolites*, which are produced after the culture has entered the stationary phase.

Industrial Products of Microorganisms

Enzymes are among the most important industrial products produced by microorganisms. Most are naturally occurring substances for which people have devised a particular use. They include amylase, used as a spot remover; pectinase, which releases fibers from flax to produce linen; proteases, used as meat tenderizers or spot removers; and streptokinases, used to dissolve blood clots. Enzymatic tools used in recombinant DNA technology, such as ligases, also come from microbes. In addition, many food additives and supplements come from microorganisms, including amino acids, vitamins, and organic acids. Microbes also produce dyes, cellulose fibers, and biodegradable plastics. Some microorganisms can convert organic materials into fuels. This process can create ethanol, methane, and hydrogen, which can be used as alternative fuels for heating, cooking, or powering vehicles. Microbes can also be used to produce antimicrobial medications; recombinant DNA technology allows microbes to produce new versions of old drugs. Genetically modified microorganisms help to produce hormones and other cell regulators. Other microbes are used in agricultural applications to protect crops from adverse conditions, such as freezing, or from pests. Bacteria or enzymes can also be combined with electronic measuring devices in **biosensors** that can detect other organisms or chemical compounds; **bioreporters** are simpler sensors composed of microbes with innate signaling capabilities.

Water Treatment

Water may become polluted *physically, chemically*, or *biologically*. In *polluted waters* the pollutant is obvious; in *contaminated water*, pollutants may not be visible. Consuming contaminated water can cause bacterial, viral, or protozoal diseases or intoxication from microbial toxins. Outbreaks are often due to *point-source infections*. In marine environments, *red tides* (eutrophic blooms of dinoflagellates) produce toxins that can contaminate shellfish.

Potable water is water that is safe to drink but may not be devoid of all microorganisms and chemicals. Water that is **polluted** is nonpotable. Permissible levels of microbes and chemicals vary from state to state. Drinking water should have a coliform count of 0 per 100 ml of water. Water is treated in three stages: **sedimentation and flocculation**, in which the addition of alum causes suspended particles to clump and settle to the bottom of the tank; **filtration**, which uses *slow* or *rapid sand filters*, *membrane filtration*, or *activated charcoal filters* to reduce the number of microbes by about 90%; and **disinfection**. Chlorine treatment is the most widely used disinfection method in the United States; it kills bacteria, fungi, algae, and protozoa but not viruses, bacterial endospores, or protozoan cysts, which must be removed by filtration. **Indicator organisms**, such as *E. coli*, are used to indicate the presence of pathogens in water. Testing methods to assess water quality include the membrane filtration method and the use of *MUG* and *ONPG* in combination with UV light.

Wastewater or sewage treatment removes or reduces contaminants in **wastewater**—the water that leaves homes or businesses after being used for washing or for flushing toilets. Effective wastewater treatment reduces the **biochemical oxygen demand** (BOD) to levels too low to support microbial growth. *Municipal* *sewage systems* collect wastewater in larger towns and cities and deliver it to treatment plants for processing, which occurs in four stages: (1) **primary treatment**, in which solids are skimmed off and heavier materials settle to the bottom as **sludge**, which is then removed; (2) **secondary treatment**, which removes pathogenic organisms and organic chemicals through *activated sludge systems* or *trickle filter systems*; (3) **chemical treatment**, in which water is disinfected, usually by chlorination, and released; and (4) **sludge treatment**, in which sludge is digested and then dried for use as landfill or fertilizer. In areas without municipal sewage treatment, **septic tanks** or **cesspools** are used to collect wastewater; solids settle to the bottom of the tank and are digested by microorganisms, while liquid flows into the surrounding soil, which acts as a filter. In agricultural areas, **oxidation lagoons** are used to treat animal waste from feedlots. Waste is pumped into a series of lagoons, in which sludge settles and is digested by microbes; then, the clarified water is released into rivers or streams. **Artificial wetlands** perform a similar purpose for some planned communities or factories, using natural processes to break down wastes through a series of ponds, marshes, and grasslands that serve to trap pollutants.

Environmental Microbiology (pp. 786–794)

Environmental microbiology is the study of microorganisms as they occur in their natural **habitats**—the location in which they live.

Microbial Ecology

The study of the interrelationships among microorganisms and the environment is called **microbial ecology**. There are different levels of microbial associations: *populations* include all the members of a single species; *guilds* are groups of microorganisms performing metabolically related processes; *communities* are sets of guilds. Within a community, populations and guilds reside in *microhabitats*. All ecosystems together make up the *biosphere*—the region of Earth inhabited by living organisms. **Biodiversity** refers to the number of species living within a given ecosystem; **biomass** is the mass of all organisms in an ecosystem. Microorganisms must adapt to changing conditions. To do this, they utilize *competition*, *antagonism*, and *cooperation*.

Bioremediation

Soil microbes break down biodegradable wastes in landfills, and methanogens degrade organic molecules to methane. Landfills are lined with clay or plastic to prevent leaching of hazardous materials into the surrounding soil and groundwater; sand and drainage pipes filter out small particles. Some *recalcitrant* substances are resistant to decay or degradation. **Bioremediation** is the use of microorganisms to clean up toxic, hazardous, or recalcitrant compounds and degrade them to harmless forms. *Natural bioremediation* involves the use of environmentally available microbes to degrade toxic substances through the addition of nutrients to stimulate growth. *Artificial bioremediation* uses genetically modified organisms to specifically degrade certain pollutants.

The Problem of Acid Mine Drainage

Acid mine drainage is a serious environmental problem resulting from exposure of certain metal ores to oxygen and the action of microbes. Rainwater leaches oxidized compounds from the soil to form sulfuric acid and iron hydroxide compounds that kill plants, fish, and other organisms and make local waterways unfit for drinking or recreational use.

The Roles of Microorganisms in Biogeochemical Cycles

Biogeochemical cycles are processes by which organisms convert elements from one form to another. These cycles include *production*, in which producers convert inorganic compounds into organic ones; *consumption*, in which consumers eat producers and other consumers; and *decomposition*, in which decomposers convert the components of dead organisms back into inorganic compounds.

- The **carbon cycle** involves the cycling of carbon in the form of organic molecules, the start of which is autotrophic. Photoautotrophic *primary producers* convert CO_2 to organic molecules through carbon fixation; chemoautotrophs accomplish the same thing using inorganic molecules as an energy source. Heterotrophs catabolize organic molecules for energy, releasing CO_2 and starting the cycle over. A growing imbalance in the carbon cycle may be involved in the retention of "greenhouse gases" and global warming.
- In the **nitrogen cycle**, microbes cycle nitrogen atoms from dead organic materials and wastes to soluble forms for use by other organisms. The process involves **nitrogen fixation**, in which gaseous nitrogen is reduced to ammonia; **ammonification**, in which amino groups from amino acids are converted to ammonia; **nitrification**, a two-step process in which autotrophic bacteria oxidize ammonium ion to nitrate; and **denitrification**, in which microorganisms oxidize nitrate to gaseous nitrogen by anaerobic respiration.
- In the **sulfur cycle**, microorganisms decompose dead organisms, releasing amino acids, which can be converted to hydrogen sulfide through sulfur *dissimulation* and then oxidized to sulfate, the most readily usable form for plants and animals. Anaerobic respiration by certain bacteria reduces sulfate back to hydrogen sulfide.
- The **phosphorus cycle** involves the movement of phosphorus from insoluble to soluble forms for use by organisms, as well as the conversion of phosphorus from organic to inorganic forms. **Eutrophication** is the overgrowth of microorganisms in waterways resulting from runoff of phosphorus and nitrogen from agricultural fertilizers; such overgrowth (*bloom*) can kill fish and other organisms.
- *Metal ions* are needed in trace amounts by organisms for growth. Metal cycling involves a transition from insoluble to soluble forms.

Soil Microbiology

Soil microbiology examines the roles played by organisms in soil, which is composed of *topsoil* rich in *humus* (organic chemicals) and *subsoil*, which is made up primarily of inorganic materials. Factors that affect the microbial populations living within soil include the amount of water, oxygen content, acidity, temperature, and nutrient availability. The majority of organisms are found in topsoil. Bacteria and fungi are both numerous in soil; soil algae live near the surface because they require light for photosynthesis; protozoa move throughout the soil but remain mostly in the topsoil. **Biomining** involves identifying and isolating natural populations of microbes that produce valuable chemicals or have useful biological functions. Most soil microbes are not human pathogens. Soilborne infections typically result from handling, ingesting, or inhaling microorganisms from contaminated feces or urine in the soil. Examples of serious soilborne infections include anthrax, histoplasmosis, and hantavirus pulmonary syndrome. Microbial plant diseases cause rot, cankers/lesions, wilt, blight, galls, growth aberrations, or bleaching.

Aquatic Microbiology

Aquatic ecosystems support fewer organisms than soil habitats because nutrients are diluted. Many organisms live in biofilms attached to surfaces that concentrate nutrients to sustain growth. *Freshwater* systems have low salt content; *marine* environments have a salt content of about 3.5%. Release of *domestic water* (water from sewage or waste treatment) affects water chemistry and microorganisms living in the water. Microorganisms distribute themselves within water systems according to oxygen availability, light, and temperature; stagnant waters are low in oxygen. Freshwater systems are made up of four zones: a **littoral zone** along the shoreline, a **limnetic zone** at the surface of the water, a **profundal zone** below the surface zone, and a **benthic zone** encompassing the deepest water and bottom sediments. Marine ecosystems are nutrient poor, and most organisms live in the littoral zone, where nutrients and light are greater. The benthic zone is the largest of the marine zones; deep ocean trenches are in the **abyssal zone**, where **hydrothermal vents** provide super-heated, nutrient-rich water to support microbial growth. Specialized aquatic environments include salt lakes, iron springs, and sulfur springs, which are inhabited by highly specialized organisms.

Biological Warfare and Bioterrorism (pp. 794–797)

Bioterrorism is the use of microbes or their toxins as *biological weapons* to terrorize human populations. **Agroterrorism** is aimed at destroying the food supply.

Assessing Microorganisms as Potential Agents of Warfare or Terror

A potential biological threat to humans is based on four criteria: *public health impact*, *delivery potential*, *public perception*, and *public health preparedness*. The same criteria can be used to assess the risk to livestock. Threats to crops are evaluated by extent of crop loss, delivery and dissemination potential, and containment potential.

Known Microbial Threats

Biological agents are categorized from highest to lowest (or unknown) weapons potential as *category A*, *category B*, and *category C agents*. Smallpox is currently considered the greatest threat to humans. Animal pathogens are similarly categorized. Foot-and-mouth disease (FMD) is considered the most dangerous potential agroterrorism agent in animals. Plant pathogens that could be used as terrorist agents are mostly fungi and could easily result in soil contamination that could impact food supplies.

Defense Against Bioterrorism

Defense involves *surveillance* combined with effective response to an attack. Category A diseases are *reportable*, allowing effective monitoring of unusual outbreaks and providing for quarantining of patients, distribution of antimicrobial drugs, or mass vaccination. Preventing agroterrorism is more difficult, because livestock are moved frequently without being tested for disease and because facilities are open to the public. Effective screening for plant and animal diseases and better diagnostic techniques, vaccines, and treatments are needed for animal pathogens.

The Roles of Recombinant Genetic Technology in Bioterrorism

Recombinant genetic technology could be used to create modified or new biological threats; scientists could also use it to identify genetic "fingerprints" to track biological agents or to develop new vaccines and treatments.

KEY THEMES

The final chapter of the text presents a snapshot of microbes in specific relationships with humans. Since Chapter 14 we have focused on the more dire results of our associations with microbes—infectious diseases. Making us sick, however, is not the only thing that microbes do, nor is it the primary thing that they do in the greater scheme of things. As you complete this initial voyage into the world of microbiology, do not lose sight of the following observations:

> *Microorganisms are responsible for the very survival of life on Earth:* Our entire ecosystem depends on the beneficial chemical reactions perpetuated by microbes every day. Microbes help to produce the food we eat and help to clean up our environment. Without microbes, nutrient cycling would cease, food chains would collapse, and humans would die.
>
> *Microbes are extremely useful for industry and research:* Their versatility and adaptability have allowed us to use them as tools to better our lives through the production of everything from pharmaceuticals to clothing dyes. As research tools, they also provide model systems for the discernment of the complexities of life. Microbes even help us to fight other microbes, making them entities to be praised, even as we fear the diseases they cause.

QUESTIONS FOR FURTHER REVIEW

Answers to these questions can be found in the answer section at the back of this study guide. Refer to the answers only after you have attempted to solve the questions on your own.

Multiple Choice

1. What is the function of microorganisms in food and beverage production?
   a. add flavor
   b. add nutritional value
   c. aid in preservation
   d. All are functions of microorganisms in food and beverage production.
2. The most common microbes used for the fermentation of vegetables are
   a. fungi.
   b. lactic acid bacteria.
   c. yeast.
   d. algae.

3. Silage is an example of fermented
   a. cabbage.
   b. milk.
   c. pork.
   d. grains and vegetables.
4. Alcoholic beverages such as wine and beer use known starter cultures for fermentation primarily to
   a. prevent spoilage, because natural microbes always spoil the product.
   b. add specific flavors natural microbes wouldn't.
   c. ensure consistency of product.
   d. speed up the fermentation process.
5. Which of the following is NOT an intrinsic factor of food that affects spoilage?
   a. water activity
   b. nutrient content
   c. physical structure of the food
   d. All are intrinsic factors.
6. Which of the following is the best example of a semiperishable food?
   a. milk
   b. cake
   c. flour
   d. ground beef
7. Which of the following food processing methods would be best to use to eliminate all potential for spoilage in foods?
   a. pasteurization
   b. canning
   c. freeze-drying
   d. gamma irradiation
8. Which food preservation technique would be best to use for long-term storage of a given food?
   a. pasteurization
   b. the use of preservatives such as salt
   c. cold storage
   d. UV irradiation
9. Which microbe is used in agriculture as a natural pesticide?
   a. *Escherichia coli*
   b. *Bacillus thuringiensis*
   c. *Pseudomonas syringae*
   d. *Saccharomyces cerevisiae*
10. During the treatment of drinking water, the majority of microbes are removed at which stage?
    a. sedimentation
    b. flocculation
    c. filtration
    d. disinfection
11. Which microbes are LEAST likely to be removed from water during treatment?
    a. bacteria
    b. fungi
    c. protozoa
    d. viruses
12. If a community didn't have standard municipal wastewater treatment, the most effective alternative would be
    a. individual septic tanks.
    b. individual cesspools.
    c. community oxidation lagoons.
    d. community-based artificial wetland.
13. Microbial biodiversity would be greatest in which of the following environments?
    a. ocean bottom
    b. farmland
    c. desert
    d. hot spring
14. Microbial diversity would be most limited in which of the following environments?
    a. ocean bottom
    b. farmland
    c. desert
    d. hot spring

15. The carbon cycle is grounded in the continual cycling of
    a. carbon dioxide.
    b. carbon monoxide.
    c. methane.
    d. organic molecules.
16. The conversion of gaseous nitrogen to ammonia is accomplished by
    a. ammonification.
    b. denitrification.
    c. nitrification.
    d. nitrogen fixation.
17. The key environmental factor affecting microbial populations living in the soil is
    a. water activity.
    b. nutrient availability.
    c. pH.
    d. temperature.
18. Natural aquatic habitats are affected most by
    a. rainwater runoff.
    b. the release of domestic water.
    c. temperature fluctuations.
    d. salt concentration.
19. Which aquatic system would be able to sustain the most microbial biodiversity?
    a. fast-flowing river
    b. deep ocean
    c. estuary
    d. salt lake
20. Of the criteria used to assess biological threats to humans, which one will most define how a community would respond to an actual attack?
    a. public health impact relating to the ability to treat victims outside the hospital
    b. delivery potential relating to the ability of the agent to spread on its own
    c. public perception relating to fear of the agent
    d. public health preparedness relating to the ability to properly diagnose the agent

Fill in the Blanks

1. Applied microbiology encompasses two fields of work and study: ___Food___ microbiology and ___Industrial___ microbiology. ___Environmental___ microbiology studies microbes in nature.
2. To maintain consistency of food products, ___cultures___ of known organisms are used for commercial fermentations.
3. Two sets of parameters affect food spoilage: ___intrinsic___ factors, which describe the food itself, and ___extrinsic___ factors, which describe how the food is handled.
4. Sugar removes ___microbes___ from foods to preserve them, and garlic produces ___allicin___ to inhibit __________________. Of the two, __________________ is the more specific antimicrobial agent.

5. "Food poisoning" can be divided into two types. Consumption of living microbes leads to ___infections___; consumption of microbial toxins leads to ___intoxications___.
6. Water that is safe to drink is called ___potable___; water that isn't safe to drink is called ___nonpotable___.
7. Compounds that cannot be degraded by natural processes are said to be ____________________.
8. The process of ____________________ utilizes microbes to clean up toxins and pollutants from the environment.
9. Individual microbes form ____________________, which in turn form ____________________ and then ____________________.
10. Microbes can interact with each other in three general ways, through ____________________, ____________________, or ____________________. Of these interactions, ____________________ is the most limiting to biodiversity.
11. The oxidation of ammonium ion to nitrate, a soluble form of nitrogen needed by plants, occurs via the process of ____________________. This process is performed only through the combined efforts of two specific genera of bacteria: ____________________ and ____________________.
12. The two major components of the sulfur cycle are ____________________ and ____________________.
13. The leaching of phosphorus and nitrogen into water systems can lead to ____________________, the overgrowth of microbes.
14. Water systems can become polluted in one of three ways, ____________________, ____________________, or ____________________.

15. The biological agents that top the list of threats to humans, animals, and plants are ____________________, ____________________, and ____________________, respectively (for plant threats, give the type of microbe rather than an actual microbe).

16. The key to defense against bioterrorism is the combination of ____________________ with ____________________; through this combination, the threat itself is reduced, and the ability to control an attack, if it occurs, increases.

Short-Answer Questions for Thought and Review

1. In the ancient world, most people generally drank wine, beer, or ale rather than water. Why might have they done this?

2. What is the major difference between food fermentations and industrial fermentations? Which type would you expect to have higher quality control requirements and why?

3. Wind, water, and solar energy have been around for years as alternatives to oil-based energy generation systems. Using microbes to produce alternative fuels is relatively new. Why would microbial fuel sources, more so than wind, water, or sun, be such an attractive option?

4. How do large-scale environmental changes, such as global warming, affect microbial communities?

5. How does the recycling of plastics, glass, paper, and other goods help to keep the carbon cycle going?

Critical Thinking

1. Home canners are more likely to experience food poisoning from the presence of botulinum toxin in their canned goods than are people who buy industrially canned goods. Why?
2. Municipal wastewater treatment in principle is, in many places, as good as water treatment in removing harmful microbes and chemicals. Why, then, isn't treated wastewater recycled back into drinking-water supplies?
3. Which method of bioremediation—natural or artificial—do you think would be the preferred method? Why?

Concept Building Questions

1. In this chapter we have learned how microbes can be used in the process of bioremediation. Given this discussion and previous materials in the textbook, describe how you could engineer microbes to fight *diseases*.
2. Given what you have learned in the textbook, why is it more likely that new, or emerging, human pathogens will continue to arise from microbes living in humans or other animal species and not from microbes living in soil or water habitats?
3. Explain why HIV could never be turned into a biological weapon. Use what you have learned in the textbook and the criteria of assessment of biological agents presented in this chapter to answer this question.

Answers to Study Guide Questions

CHAPTER 1 A Brief History of Microbiology

Multiple Choice

1. a	3. c	5. c	7. b	9. c
2. d	4. d	6. a	8. b	10. d

Fill in the Blanks

1. microbes
2. abiogenesis
3. pathogen
4. nosocomial
5. Edward Jenner

Short-Answer Questions for Thought and Review

1. Viruses do not have the ability to replicate, transcribe, or translate on their own; they have no metabolism; they cannot exist independently. Prokaryotes and eukaryotes can do all of these things.
2. To some extent, what has changed is that we want more detailed information regarding the inner workings of the cell. We are interested in the specific processes of metabolism and the specific functioning of genes, both of which were not well understood, if even proposed, during the Golden Age of microbiology. What has stayed the same is our desire to understand infectious diseases and how to control them.
3. Control groups are necessary to the scientific method because they give reference points to show when a reaction has actually occurred. If Pasteur had not shown that his open swan-necked flasks could remain clear for months on their own before the introduction of dust, it would have been impossible to demonstrate that exposure to actual microbes was necessary to seed the flasks and allow more microbes to appear.
4. Both poor food and unsanitary conditions weaken individuals, making them more susceptible to becoming sick (lowers immunity). Unsanitary conditions themselves place individuals at risk by continually exposing them to agents that could make them sick—eventually, with enough exposure, you cannot avoid picking something up.

Critical Thinking

1. Organisms related by genetics and environment can be expected to behave the same in similar situations. By grouping microbes with appropriate methods of classification, scientists can make predictions of how microbes cause disease. If something is known about how one microbe in a group causes disease, then most likely similar conclusions can be drawn for related species. One drug could be useful against many, and vaccines might be useful for preventing several diseases.

2. The scientific method is based on proposing a hypothesis, designing experiments to test the hypothesis, drawing conclusions, and refining questions. Semmelweis could not have designed an experiment to test his hypothesis that physicians were transmitting disease because he could not intentionally infect humans with disease. Washing hands was shown to reduce disease transmission and the observation alone was sufficient to support his hypothesis, without having to allow purposeful infection to occur.
3. It is unlikely that a single "magic bullet" will be found to cure all infectious diseases because so many diverse agents cause disease. Diversity implies a lack of conformity when it comes to responding to chemical agents designed to kill, and it is very unlikely that any one chemical/drug/vaccine, etc., will ever be found that can react against all microbes and viruses.

Chapter 2 The Chemistry of Microbiology

Multiple Choice

1. d	4. c	7. b	10. b
2. a	5. c	8. a	11. c
3. c	6. d	9. a	12. c

Fill in the Blanks

1. protons
2. atoms, molecule
3. O=O, O::O
4. Reaction a: decomposition reaction; Reaction b: exchange or transfer reaction; Reaction c: synthesis reaction
5. hydrogen (H^+), hydroxyl (OH^-)
6. buffers, protein
7. hydrophilic, hydrophobic
8. peptide, hydrogen
9. deoxyribose, ribose
10. DNA sequence: 3' — TAACGATGGCTA — 5';
    RNA sequence: 3' — UAACGAUGGCUA — 5'

Short-Answer Questions for Thought and Review

1. An element (gold is an example) is matter composed of only one type of atom. A molecule (molecular oxygen, O_2, is an example) is matter composed of two or more atoms held together by chemical bonds. A compound (water, H_2O, is an example) is matter composed of two or more different elements.
2. See page 30 of Chapter 2 in the textbook for pictures of the electron shells of both calcium and chlorine. The valence of calcium is +2 and the valence of chlorine is –1 (thus 2 chlorine atoms are needed to balance the +2 of calcium). Chlorine has 7 electrons in its valence shell and calcium has 2; each chlorine will pull one of calcium's electrons to it to complete the shell.
3. Polyunsaturated fats would contribute least to atherosclerosis because they will not pack as well and should not produce the solid deposits in the

arteries that cause heart disease (they should remain more liquid at body temperature).

4. The five primary functions of proteins are: (1) structure, (2) catalysis, (3) regulation, (4) transportation, and (5) defense and offense. Without proteins to give the cell structure, the cells could not physically form or maintain stable shape in the environment. Without proteins to perform catalysis, metabolism would not occur and the cell would die. Without proteins to regulate the function of the cell, it could not know when to take in nutrients, grow, or divide, and thus would not be able to respond to its environment. Without proteins for transportation, the cells could not move molecules across their membranes and thus could not acquire food. Without proteins for defense and offense, the cell would be unable to protect itself from pathogens.

Critical Thinking

1. Metabolism is the sum total of all chemical reactions inside an organism. Anabolism is the reaction that builds and catabolism is the reaction that takes apart (synthesis and decomposition, respectively). The bacterium in the human intestinal tract would be expected to have higher metabolic function because it has more nutrients available to fuel catabolism, which in turn drives anabolism of new components or microbes. Fewer nutrients will be available to the bacterium living in the flowerpot.
2. Microbes that can drastically alter the environment to suit their own needs can outcompete other microbes essentially by eliminating them. Organisms that cannot adapt to acid conditions, for example, will die, leaving the acid producer in charge of the field. Specialization of environment is one way microbes survive and is one of the things that makes them so prevalent in the biosphere.
3. Neutral mutations predominate in nature because they run the least risk of doing something harmful. If a protein is destroyed by a single substitution, then chances are some function of the cell no longer works. That cell is less able to survive and thus dies before propagating. It is, essentially, evolutionarily deselected. Mutations that have no particular effect on survival do not get deselected and thus remain in the population.

Concept Building Questions

1. In the fermentation of wine, sugar molecules are broken down into smaller alcohol molecules as well as carbon dioxide. The chemical reaction represented is a decomposition reaction. Decomposition reactions are important because they allow microbes to take large, complex molecules and break them down into smaller, more versatile molecules that can be used in a variety of subsequent chemical reactions.
2. Understanding how chemical bonds form helps us to understand how relationships form between molecules and organisms in that it gives us a sense of permanence (covalent vs. ionic associations), specificity (more bonds/better bonding usually implies a closer "match" between objects), and coexistence possibilities (the longer microbes live together, the more probable it is for them to engage in shared associations). In the environment, these ideas can tell us which microbes live together, in a certain place, how long they have lived together, what types of animals a pathogenic organism can infect, etc.

Chapter 3 Cell Structure and Function

Multiple Choice

1. d	5. d	9. c	13. b	17. c
2. b	6. b	10. d	14. c	18. d
3. a	7. a	11. a	15. b	19. d
4. d	8. b	12. c	16. d	20. c

Fill in the Blanks

1. capsule, slime layers
2. flagellin, tubulin
3. N-acetylglucosamine (NAG), N-acetylmuramic acid (NAM)
4. Lipid A, LPS
5. hopanoids
6. Part a: moves into the cell; Part b: does not move; Part c: moves out of the cell, moves into the cell, moves out of the cell, moves into the cell, moves out of the cell
7. isotonic, out of, shrink
8. the same, different, active
9. smaller, 70, 30, 50
10. cellulose, chitin, glucomannan
11. cholesterol
12. microtubules, microfilaments, intermediate filaments
13. secretory vesicles, exocytosis
14. nucleus, mitochondrion
15. mitochondria, chloroplasts

Matching

1. F	3. H	5. B	7. C	9. E
2. J	4. I	6. G	8. A	10. D

Short-Answer Questions for Thought and Review

1. Movement: flagellum (prokaryote only is shown), cilia (eukaryotic only); protection/adherence/interface with outside of cell: cell membrane, cell wall, glycocalyx; *control center*: nucleoid, nucleus (nuclear envelope, perinuclear space, nuclear pore, nucleolus); physical structure: cell wall, cytoskeleton; "mechanical operation of cell": cytoplasm, ribosomes, lysosomes, mitochondria, centriole, secretory vesicle, Golgi body, transport vesicle, rough and smooth endoplasmic reticulum; storage: inclusions (prokaryotic only)
2. Bacteria use flagella to move by runs and tumbles. To move toward an attractant, runs must be more frequent than tumbles and be longer, to get the organism moving in the general direction of the attractant.
3. Corkscrew motility is caused by the axial filament, a flagellum wrapped around the body of the microbe. It might help invasion because the corkscrew motion is one designed to embed something into a surface (think about a screw being turned into a piece of wood). The end of the microbe will contact the tissue surface and as it moves and works its way in, the corkscrew form will hold it in place.

4. Gram-positive infections are "less damaging" because they do not produce an endotoxin. Thus, when the organisms die, there is no secondary toxic effect from the components of the cell wall being released into the host. In Gram negatives, when the cells die, LPS, containing lipid A, is released and causes immunological response damage.
5. Essentially, the diagram will look like part c of Figure 3.19. The movement of protons coupled to an ATPase provides the energy for the symport of glucose into the cell.
6. Because of the differences between prokaryotes and eukaryotes in terms of structure and function, it makes sense to target the differences when using antimicrobial drugs. This way the drugs kill the bacteria and leave host tissues alone because they can't recognize or react with them.
7. The Gram-positive cell wall presents a rigid barrier that does not allow the extrusion of cytoplasm to form pseudopodia; thus the cells can't phagocytize anything.

Critical Thinking

1. Metabolism is probably the most important determinant of life because it is the one thing that ensures independent survival. Organisms must be able to take in nutrients, process them, use them, excrete wastes, and repeat the process simply to maintain cohesion, function, etc. A chair cannot metabolize and cannot continue to "survive" unless someone takes care of it, polishes it, and fixes it when it is broken. It does not cease to be a chair because it cannot communicate or respond to the environment.
2. Organelles allow for compartmentalization and a higher degree of organization within the eukaryotic cell than in the prokaryotic cell. Organization allows for more complexity because there is a way to regulate what is going on. Thus, organelles allow for the grouping of like functions and linking of functions across the cell; therefore the cells can afford to be larger and more complex.
3. Glucose 6-phosphate is physically and chemically different from glucose. Glucose, therefore, will be "high" outside the cell but "low" (to nonexistent) inside the cell because it does not exist as glucose once it is translocated. Glucose will therefore continue to be drawn into the cell because the gradient will always be high to low (outside to inside).
4. You would look most for the remnants of metabolic processes that would indicate some ability to survive independently.

Concept Building Questions

1. The hydrophilic head groups will be held together by weak ionic bonds and some hydrogen bonds; the hydrophobic tails will associate hydrophobically (they exclude water; the carbon and hydrogen atoms themselves are held together by nonpolar covalent bonds). Because the only true bonding is occurring with the hydrophilic region, high temperatures, which increase movement and ultimately will disrupt noncovalent bonds, will have a tendency to unzip the bilayer (there are no bonds internally holding one side to the other). Thus, a single layer is far more stable because it can't "melt."
2. Leeuwenhoek was the first to describe bacteria, though he didn't know what they were. He would have been able to see inside eukaryotic cells, but most likely not inside prokaryotes. He would have noticed a size difference, differences in overall shape, and, depending on how he prepared his samples, he

might have been able to see flagella or capsules or other external structures. He would have been able to see some of the internal architecture of eukaryotic cells, most notably the nucleus, and may have been able to deduce that his wee "beasties," being so much smaller, probably didn't have one.

Chapter 4 Microscopy, Staining, and Classification

Multiple Choice

1. d	4. d	7. b	10. b	13. d
2. c	5. d	8. c	11. d	14. a
3. b	6. c	9. a	12. d	15. a

Fill in the Blanks

1. Part a: 5×10^{-6} mm; Part b: 2.5×10^{5} µm; Part c: 6 nm
2. contrast, resolution
3. better, shorter
4. differential interference, scanning electron
5. ultrastructure
6. contrast, resolution
7. acidic, basic, basic, acidic
8. counterstain
9. classification, nomenclature, identification
10. agglutination, antibodies, antigens

Short-Answer Questions for Thought and Review

1. See page 103; use fluorescent microscopy and the dye fluorescein isothiocyanate that binds specifically to anthrax bacteria and makes them glow green.
2. Artifacts can distort the appearance of microbial cells or cause them to stain oddly. Both of these problems could make the microbe look enough like another microbe to make identification difficult.
3. Simple stains are used to gain general morphological information about a specimen, such as general shape and arrangement. This by itself, however, is not particularly discriminating when looking at many microbes. Differential stains, which are specifically designed to give "either/or" -type answers, allow you to begin to categorize microbes. They are thus more useful.
4. The three domains are Eukarya, Bacteria, and Archaea. Eukarya contains animals, plants, fungi, and protists. Prokaryotae was split into Bacteria and Archaea.
5. In order of most specific to least specific, the tests are: nucleic acid analysis, serology, biochemistry, and morphology.

Critical Thinking

1. You should always use lowest power (4×). At low magnification you are presented with a larger perceived field of view and smaller specimens; this allows you to rapidly scan wide areas of a slide to find what you want to examine more closely. Using a higher power cuts down on the field of view and makes finding specimens very difficult.

2. Most diagnostic tools are composed of generally less precise methods because they are usually cheaper, easier, and faster to perform than more specific methodologies. In a diagnostic situation, you may not have the time to wait for a definitive identification and in many cases groups of organisms can be treated with general classes of drugs, making a more general identification appropriate and useful.
3. The organism is *Corynebacterium pseudodiphtheriticum*.
4. What the key and the chapter should suggest is that no one test is sufficient to clearly identify any microbial species. All Gram-positive rods, for example, essentially look similar and thus cannot be conclusively identified based on appearance. Doing only a single metabolic test (acid from mannitol) also would not be distinctive, as many organisms make acid and many don't. Using several tests in combination, however, increases the chances that a distinction can be made. Observation is the basis of the keys, followed by biochemistry. By doing several tests starting with morphology, choices can be eliminated and a path determined that will lead to a proper, logically determined identification.

Concept Building Questions

1. The answers are as follows:
   a. If you look at the last row of Figure 4.22, there isn't much difference in appearance between the three ticks shown. While an expert might be able to tell the difference, most clinicians, laboratory workers, and average individuals would not be able to tell the difference based only on appearance. This lack of clear distinction in appearance can lead to mistakes in classification. Evolutionarily, just because two natural objects look similar doesn't mean they took the same path to get there.
   b. Based on morphology, the presence/absence of a nucleus, the presence/absence of cilia, the presence/absence of fimbriae/pili, the presence/absence of membrane-bound organelles, and the structure of the DNA, appearance could be used to differentiate between prokaryotes and eukaryotes. Size, the glycocalyx, the cell wall, the cytoplasmic membrane, inclusions, and ribosomes would not be distinctive enough between the two groups to allow definitive identification.
   c. At a very gross level, genera could be distinguished by the presence/absence of things like a glycocalyx, flagella, fimbriae, pili, and cell wall type, but all species in a genera will essentially share the characteristics of that genera. Species are more distinguishable based on things that can't be seen, such as metabolism or genetics.

Chapter 5 Microbial Metabolism

Multiple Choice

1. b
2. d
3. d
4. b
5. d
6. b
7. d
8. d
9. c
10. a
11. c
12. d
13. b
14. c
15. b
16. b
17. c
18. b
19. b
20. c
21. b
22. b
23. c
24. d
25. b
26. d
27. a

Fill in the Blanks

1. precursor metabolites, energy, exergonic, release
2. donors, oxidation, acceptor, reduction
3. NAD^+, $NADP^+$, FAD
4. competitive, noncompetitive, competitive
5. respiration, fermentation
6. pentose phosphate pathway, Entner-Dourdoroff pathway, pyruvic acid
7. answers are as follows: (1) Glycolysis: cytosol, cytosol; (2) Krebs cycle: cytosol, mitochonrial matrix; (3) Electron transport chain: cytoplasmic membrane, inner mitochondrial membrane; and (4) Photosynthesis: cytoplasmic membrane invaginations called thylakoids, invaginations of the inner chloroplast membrane which form thylakoids
8. protons, proton gradient, chemiosmosis
9. oxygen, inorganic molecules
10. proteases, deamination, Krebs cycle
11. Cyclic, PS I, Noncyclic, PS I, PS II
12. amphibolic

Matching

1. A, B, C, D, G
2. A, D, F
3. A, E
4. F
5. B, C
6. D, G
7. B, E
8. B
9. A, C, G
10. A, B, C, D, G

Short-Answer Questions for Thought and Review

1. Allosteric inhibition is a type of noncompetitive inhibition in which regulatory molecules bind to enzymes at sites other than the active site and turn the enzyme off. The binding is irreversible and generally involves changing the shape of the enzyme so that it can no longer function (cannot bind substrate). Allosteric inhibition essentially "removes" enzymes from the mix; thus, the only way to overcome the inhibition is to produce more enzyme above the level of the inhibitor.
2. Fermentation is necessary to regenerate electron carriers (i.e., NAD^+) to keep glycolysis running. Cells that ferment can't use cellular respiration to do this and so the only way to keep making even small amounts of energy is to keep using glycolysis, which means continually recycling the supply of NAD^+.
3. Catabolism and anabolism are essentially two halves of the same coin. Catabolism breaks down molecules into pieces that anabolism uses to construct new macromolecules for the cell. If anabolism proceeded at a faster rate than catabolism, eventually the anabolic systems would run out of building blocks and the cell would stop synthesis. If the cell were attempting to divide, it wouldn't be able to because it couldn't produce all of the new parts it needed.
4. An example is given for step 3 of the Krebs cycle as shown in Figure 5.19. NAD^+ is the oxidized form of the electron carrier; NADH is the reduced form. Isocitric acid and a-ketoglutaric acid are both Krebs cycle intermediates. In the equation below, the electron donor is the isocitric acid that loses COOH in its enzymatic conversion to α-ketoglutaric acid. The

hydrogen atom is transferred to NAD^+, the electron acceptor, which becomes reduced in the process. CO_2 is evolved off essentially as waste. Overall, the conversion of isocitric acid to α-ketoglutaric acid is an oxidation reaction and the conversion of NAD^+ to NADH is a reduction reaction.

$$\text{Isocitric acid} + NAD^+ \rightarrow \alpha\text{-ketoglutaric acid} + NADH + CO_2$$

Critical Thinking

1. The advantage to synthesizing everything you need from a small number of precursors is that you are a bit less dependent on finding correct nutritional equivalents in the environment. Such an organism can live in an environment where nutrients are not readily available or where they appear intermittently. The disadvantages would include the fact that the organism will always be using a lot of energy to synthesize everything, and to some extent will have a higher burden to keep many more metabolic pathways functioning normally. It's best to be able to do both because when nutrients are available, the organism can save some energy by importing materials rather than making them, but when the nutrients aren't there it won't starve because it can make what it needs. In this way, it can survive the variability of the environment to a much greater extent than microbes that cannot switch metabolic functions.
2. There are a couple of ways the cell can control amphibolic pathways. In the case of glycolysis and gluconeogenesis, the key points where the pathway would become dedicated to catabolism or anabolism, respectively, are actually controlled by different enzymes. Thus if you want glucose, a different set of enzymes is activated than if you want to break down glucose. The pathways are controlled at the point of entry. Another way such pathways can be controlled is through the use of inhibitors and/or feedback-type situations in which the products of the pathway (both catabolic and anabolic) can be used to modulate the activity of enzymes in the pathway to direct the pathway in one direction or another depending on need.
3. In the oxygenated environment, cellular respiration using the electron transport chain will be the dominant energy generation mechanism. ATP is made primarily from the proton motive force generated by the electron transport chain. If the organism is suddenly moved to an anaerobic environment, respiration will stop once the last terminal electron acceptor (oxygen) is used up. At this point, one of two things will happen. If the cell is capable of fermentation, it can continue to generate very small amounts of ATP from the continual cycling of glycolysis, using fermentation to regenerate NAD^+. If the cell cannot ferment, and cannot use an alternate electron acceptor, then it will most likely die.

Concept Building Questions

1. Microbes can't import whole proteins inside rather than make them because proteins are simply too large to pass through the outer structures microbes possess. Prokaryotes have complex cell walls that are breached only by special channels or pores. Other microbes have cytoplasmic membranes at the very least (some will also have other structures) that are also impassable by large proteins. Proteins cannot diffuse through a membrane, nor are the channels, pores, and other transporters large enough to allow such big molecules through. If they

were, the holes would be so big in the membranes of microbes that everything else could pass in and out of the cell with little control and most likely the cell would burst because of its inability to control movement of small molecules.

2. Enzyme-substrate complexes will be formed via weak interactions between the two: hydrogen bonds, electrostatic associations, and some conformationally derived interactions. Essentially, those bonds that can be easily formed and re-formed will be used in the enzyme-substrate complex. Covalent bonds of any sort cannot be used to establish this complex because they are not readily dispensable. A given enzyme may be needed to break a covalent bond in a substrate, but it is not going to be able to work on the substrate to the extent it needs to do that if it is itself permanently bound to the substrate. Thus, covalent associations would prevent the enzyme from moving the substrate, aligning it in the correct orientation, etc. In other words, function would cease even though the enzyme would be bound to the substrate.

Chapter 6 Microbial Nutrition and Growth

Multiple Choice

1. d
2. b
3. a
4. c
5. c
6. d
7. b
8. d
9. d
10. d
11. b
12. d
13. a
14. d
15. b
16. b
17. c
18. a

Fill in the Blanks

1. nutrients
2. do, *E. coli*
3. nitrogen fixation
4. Psychrophiles, 45
5. acidophiles
6. broth, colonies
7. streak plate
8. Enrichment
9. 6300+
10. Turbidity, indirect

Short-Answer Questions for Thought and Review

1. A limiting nutrient is one that can inhibit metabolism if it is not present. Hydrogen is not a limiting nutrient. It is the most common chemical element in cells, and metabolism is never interrupted by a lack of hydrogen.
2. Aerotolerant and microaerophilic microbes have some of the enzymes needed to protect them from reactive oxygen species, but either not all of them are present or they are present in insufficient amounts to be completely protective against atmospheric levels of oxygen. Low oxygen levels, however, can be accommodated.
3. Tolerant is better because it means you can survive with or without the environmental factor being there. If there is no acid, you are just as capable

of surviving as if there are moderate levels of acid. Obligates are much more restricted in that they actively require the condition to be there (and will die if it is not).

4. Antacids make the digestive tract more alkaline and *H. pylori* prefers alkaline conditions. Thus you are creating a favorable environment for their growth.
5. Contamination can occur during collection (microbes from other body areas), during transport (incorrect storage or contaminated supplies), or during inoculation of media at the lab (poor aseptic technique).

Critical Thinking

1. Carbon is a structural element as well as a nutritional element used to run metabolism. It is therefore needed in great quantities because it is used so much. Iron is used primarily in electron transport chains; it is critical to survival but present only in a small subset of proteins in the overall cell.
2. Single-celled microbes, such as bacteria, can survive over a broader temperature range than multicellular organisms. Bacteria can survive at temperatures well below their optimal growth temperature, because their metabolic activity slows drastically. Additionally, at these low temperatures the integrity of the cell wall is not altered due to disrupted hydrogen bonds and denatured proteins.
3. Fastidious organisms require many specific nutrients at specific levels. Few places in the environment are so specific. A human body, however, would be a perfect home for a fastidious organism, because we ourselves require nutrients in very specific levels.

Concept Building Questions

1. The most obvious effect of limiting nutrients is a decrease in metabolism; without nutrients to bring into the cell, there is nothing to fuel catabolism and without catabolism, anabolism does not occur (thus no metabolism). A more subtle effect is seen with transport mechanisms. As one nutrient becomes limiting, many microbes have the ability to use others. This, however, generally requires the switch from one type of transporter to another. Thus, the flux in nutrients requires the cell to make many different changes to the proteins it expresses. This ultimately uses energy, and puts a bigger demand on metabolism to produce it.
2. The solution added to lyse the red blood cells should have a lower concentration of NaCl. This would be a hypotonic solution. This will cause water to move into the red blood cells. The result is that the cells will swell due to the increase in fluid and burst.
3. Many colonies can have similar morphologies yet not be related. Though some bacteria produce very distinctive colonies that are readily identifiable, most don't, and so morphology alone cannot be used for the majority of microbes. As we learned in Chapter 4, several tests are generally necessary to truly establish the identity of an organism. In this case, colony morphology coupled with microscopy and metabolic tests would be far more useful than colony morphology alone.

Chapter 7 Microbial Genetics

Multiple Choice

1. b
2. d
3. b
4. a
5. b
6. c
7. d
8. d
9. b
10. c
11. c
12. a
13. b
14. c
15. b
16. d
17. d
18. d
19. b
20. b
21. d
22. c
23. d
24. c
25. a

Fill in the Blanks

1. uracil, thymine
2. linear, more
3. triphosphate deoxyribonucleotide
4. leading, lagging
5. transcription, translation
6. self-terminating, enzyme dependent
7. codons
8. mRNA, tRNA, rRNA, translation
9. specific, more than one
10. A, P, E
11. promoter, operator, one or more genes
12. inducible, repressible
13. inducible
14. base-excision repair
15. auxotrophs
16. homologues
17. transformation
18. F plasmid, pilus (conjugation pilus)

Short-Answer Questions for Thought and Review

1. Plasmids carry "extra" genes that are not essential to survival, but which can enhance growth potential and in some cases confer pathogenicity upon the microbes that possess them. Resistance plasmids, for example, carry genes coding for antibiotic resistance (enzymes to degrade antibiotics). Virulence plasmids code for a variety of proteins, enzymes, structures, or toxins that influence pathogenicity. F plasmids, though present to stimulate conjugation, can lead to the inadvertent transfer of antibiotic resistance or other harmful genetic elements.
2. The genotype is not entirely translated into phenotype. Cells in general have many more genes than they have expressed proteins. Additionally, not all genes that are used are used at the same time, so while one set of genes may be producing a given phenotype, expression of a different set of genes may give you something different (with the two sets never being expressed at the same time).
3. Eukaryotic mRNA is first made as pre-mRNA, which has to be spliced to remove introns and leave exons; it codes for only one polypeptide at a time; and it is not translated until fully transcribed and transported out of the nucleus. Prokaryotic mRNA does not have introns; can code for more than one protein, in some cases; and can be translated as it is being transcribed.

4. Wobble allows for redundancy in the genetic code, which cuts down on mistakes in translation. If several codons call for the same amino acid and differ only in the last slot, it is highly probable that the right amino acid will be used in the majority of cases. This protects against mutational changes to the proteins produced. Wobble cuts down on energy because the cell does not have to make a separate, specific tRNA for every single amino acid codon combination possible (per the text, rather than making 64 or so codons, *E. coli* can make 40 and get the same level of translation).
5. In prokaryotes, replication, transcription, and translation all occur in the cytoplasm of the cell (though replication and transcription will be more localized to the nucleoid region). In eukaryotes, replication and transcription both occur in the nucleus, and translation occurs in the cytoplasm or bound to the endoplasmic reticulum. Some of these functions also occur in the mitochondria and chloroplasts.
6. Rates of mutation: $X = [(6 - 5) / 5] \times 100 = 20\%$, $Y = 460\%$, $Z = 200\%$. Order of increasing mutagenic potential = X, Z, Y.
7. The genetic assumption of the Ames test is that all DNA is essentially the same structurally and functionally within all cells. Thus, if something damages DNA in one cell type, there is every reason to believe it will damage DNA in another cell type.

Critical Thinking

1. RNA viruses have a higher mutational rate because they have nothing to compare their new genomic strands to in order to see if there are any mistakes. With dsDNA there is always a second strand paired to the first; if pairing can't occur, it is a signal that a mistake has been made and it can be repaired. With single strands, there is no check template; thus mutations are not caught based on structural anomalies and can slip through.
2. The mRNA sequence is 5' — AUG GCC GAU GCU GCC UCC CUA CUC UUU GUA — 3'; the protein sequence is NH_2 — f-Met — Ala — Asp — Ala — Ala — Ser — Leu — Leu — Phe — Val — COOH.
3. The RNA transcript is 5' — CGU GAU AGA CGC GUA — 3'; if you substitute the first G with an A you get a start codon: 5' — C **A**UG AUA GAC GCG UA— 3' (protein would start at AUG and skip the first C and last UA to give f-Met — Ile — Asp — Ala); if you insert an A after the first G you also get a start codon: 5' — CG **A**UG AUA GAC GCG UA — 3' (protein would start at the AUG, skip the first CG and last UA, and give the same protein as above); if you delete the second A in the sequence you will get a start codon: 5' — CGUG AUG ACG CGU A — 3' (protein would start at the AUG, skip the first CGUG and the last A, and give f-Met — Thr — Arg).

Concept Building Questions

1. The actual genes themselves and the DNA composing them never directly interact with the environment in any functional way. DNA may be present outside of the cell, but it is inert unless picked up by another cell during some sort of horizontal gene transfer process. Structure, however, interacts every day with the environment and will function accordingly. Metabolism is dependent on pulling in nutrients and excreting wastes. Thus it is the physical part of the cell and the functional parts of the cell

(the *products* of the genes) that interface with "the world," determine associations and successful reproduction, and ultimately form the basis of evolutionary selection.

2. In the presence of galactose, the cell produces a symporter and the catabolic genes needed to import galactose and break it down. Galactose is probably serving as an inducer for an inducible operon. Galactose keeps the operon on and prevents repression. If galactose runs out, there is no more galactose to bring into the cell and eventually all galactose inside the cell is used up. The repressor is now free to bind to the operator of the operon to shut off synthesis of transporters and enzymes. No new transporters are placed on the cell surface and the ones that are there can be retrieved into the cell by endocytosis and recycled along with the enzymes. A very few symporters are probably left on the cell surface to "sense" for more galactose. If galactose shows up again, it can be brought into the cell, can interact with the repressor, and can reinitiate promotion of the operon.
3. Hydrophobic amino acids will not form hydrogen bonds with water in an aqueous environment and thus will not be stable in the presence of water. One of two things could happen to the mutant protein: either the amino acid could distort the protein by folding inward to "hide" from the water, or two proteins could associate at the hydrophobic residues to "hide" each other, thus forming aggregates. In both cases, the protein will no longer interact normally with other proteins or molecules in the environment. If it is distorted, normal binding sites will most likely have been lost, and if it is aggregated, there is a good chance important interaction sites will be blocked.

Chapter 8 Recombinant DNA Technology

Multiple Choice

1. d
2. c
3. b
4. c
5. b
6. b
7. c

Fill in the Blanks

1. introns
2. sticky, blunt, sticky
3. PCR (polymerase chain reaction)
4. size, shape, charge
5. DNA, RNA
6. DNA fingerprinting
7. transgenic

Short-Answer Questions for Thought and Review

1. The three major goals of recombinant DNA technology are to get rid of undesirable traits in various organisms, engineer beneficial traits into other organisms, and create new organisms. The ethical controversy centers around the question of whether we *should* do any of this, whether we have the *right* to do this, and what will *happen* if we do.
2. The central medical focus of recombinant DNA technology is to fix what is genetically wrong in humans (cure diseases) or to improve the life of those

who are ill (i.e., produce new therapeutic drugs). In agriculture, the focus is on making hardier plants that survive better, resist disease and pests, and are more nutritious. Both venues seek to improve the organism in question, but for somewhat different purposes (health vs. productivity).

3. Three benefits of recombinant DNA technology are: (1) gene therapy to cure diseases, (2) better drugs, and (3) better research tools. Three potential hazards are: (1) unexpected environmental consequences from the release of genetically engineered microbes, (2) unwanted deaths from gene therapy, and (3) unexpected health hazards due to changes in the nutritional quality of foods.

Critical Thinking

1. The possible sequences are listed below. The left hand column shows the mRNA, and the right hand column shows the cDNA.

mRNA Sequence	*cDNA Sequence*
5' — AUG GCU UUU — 3'	5' — AAA AGC CAT — 3'
5' — AUG GCU UUC — 3'	5' — GAA AGC CAT — 3'
5' — AUG GCC UUU — 3'	5' — AAA GGC CAT — 3'
5' — AUG GCC UUC — 3'	5' — GAA GGC CAT — 3'
5' — AUG GCA UUU — 3'	5' — AAA TGC CAT — 3'
5' — AUG GCA UUC — 3'	5' — GAA TGC CAT — 3'
5' — AUG GCG UUU — 3''	5' — AAA CGC CAT — 3'
5' — AUG GCG UUC — 3'	5' — GAA CGC CAT — 3'

Concept Building Questions

1. Answers are as follows:
   a. Codon wobble, because it introduces uncertainty in the third position of the codon, can potentially decrease your ability to "match" sequences. If you are comparing sequences from two organisms for the purposes of determining relationships based on DNA, you have to ask yourself whether any differences you see between the two DNAs are due to wobble or to actual differences between the species. On long sequences, this usually becomes less of a problem because the degree of similarity will be more evident.
   b. There probably is enough DNA in the sequence, but scientists usually like to look at much more than a tripeptide to confirm relatedness. The probability that any one small stretch of DNA is similar to another is too high to base a judgment on from something so short. It is less likely, over long sequences, that similarity is due to random chance. Thus your ability to relate organisms becomes greater the longer the sequence you have to compare.
   c. GC pairs share three bonds between them, whereas AT pairs share only two bonds. Therefore, sequences rich in G + C are held together more tightly than sequences rich in A + T. The sequences presented are really

too short to get a full sense of the actual sequence, but of the sequences given, those with 5 GCs (corresponding to rows 4, 6, and 8) will be a bit stronger than those with fewer. The sequences with lower stability would therefore be those with only 3 GCs (corresponding to rows 1 and 5).

Chapter 9 Controlling Microbial Growth in the Environment

Multiple Choice

1. b
2. d
3. a
4. a
5. c
6. b
7. b
8. c
9. b
10. d
11. c
12. c
13. d
14. c
15. a

Fill in the Blanks

1. disinfectant
2. bacteriocidal, bacteriostatic
3. increases
4. Thermal death point
5. Moist, dry
6. Thermoduric, thermophilic
7. particulate, electromagnetic
8. phenolic disinfectant
9. tincture, iodophor
10. Antibiotics

Short-Answer Questions for Thought and Review

1. From least effective to most effective, the techniques are: degerming, disinfection, sanitization, pasteurization, and sterilization.
2. The two modes of action are: (1) disrupt the integrity of the cell (physical structure), and (2) disrupt the functions of metabolism or reproduction. For the first point, damage is done to the cell wall and/or cell membrane so that physical integrity is lost; this leads to a loss of stability, osmotic imbalance, and eventually lysis. For the second point, damage is done to proteins or nucleic acids (thus damage is done to structural elements), but it isn't the structural damage but rather the corresponding loss of functional capacity that is the problem. If the cell can't metabolize, it can't survive, cannot reproduce, and ultimately dies.
3. Highly effective chemical control agents (oxidizing agents, aldehydes, and gaseous agents) are smaller molecules that will be able to work their way into the interior of cells and cause damage. Because they can penetrate, they will be effective against more cell types. They are also more destructive to proteins than other chemicals and can damage enough protein content to kill or inhibit a high proportion of microbial cells.

Critical Thinking

1. The in-use test would probably be the most suitable. It allows for testing of disinfectants within the environmental setting that they would actually be used in to identify the ones that are most effective. In this test, swabs are

taken from objects before and after application of a disinfectant. The swabs are then incubated and examined for microbial growth. This test allows for a more accurate determination of the disinfectants and their strengths that are appropriate for a given situation.

2. The methods that inhibit growth of microbes by removing water are: (1) desiccation, which removes water from cells by drying (sunlight, hot air); (2) lyophilization, which involves freezing cells and then using a vacuum to remove frozen water by sublimation; and (3) osmotic pressure, which uses sugars and salts in hypertonic solutions to draw water out of cells, thereby drying them. In all cases the loss of water prevents metabolism from occurring.
3. In this case the CDC is erring on the side of caution. There is no proof that triclosan increases resistance, but there is also no firm proof that it does not. As degerming with normal, nontriclosan soaps is just as effective if done properly, there is no detriment to restricting the use of antimicrobial soaps and thus no reason not to be cautious. In the event that triclosan does prove to increase resistance, having already begun to restrict use will decrease the effort required to get triclosan out of the environment.

Concept Building Questions

1. The fundamental structural difference between the microbes most susceptible to antimicrobial agents and those most resistant comes down to the external coverings of the cells. Endospores, cysts, and encapsulated organisms are most resistant because they have very thick, impermeable protective layers between them and the environment. Nothing can travel through these layers very well and thus chemical agents won't be able to do much damage. For the most susceptible, no such outer coverings exist (the Gram-positive cell wall is thick, but not nearly as impermeable or protective as spore walls) and therefore antimicrobial agents can penetrate better. To be an effective germicide and kill the resistant organisms, the germicide must have penetrating power—by some mechanism it must bypass or physically destroy the protective outer layers to expose the organism underneath.
2. A microbe's form of resistance will differ depending on the organism and the chemical agent in question. Only a few of many possibilities are therefore presented.
   a. If an agent inhibits growth by bonding to a protein and inhibiting some function, then natural mutations in the protein that prevent bonding will prevent agent activity. If an agent targets formation of a structure, then changes to the structural elements such that they can't interact with the agent will prevent activity.
   b. If an agent targets function, such as metabolism, then alternate pathways that bypass the inhibited spot can circumvent agent activity. Alternatively, induction of a transporter that can pump the agent back out of the cell can result in not enough of the agent remaining inside to cause damage.
   c. If the agent binds DNA to prevent transcription (or otherwise modifies DNA so that transcription cannot occur) then changes to the DNA such that the agent can't alter it will result in loss of function for the agent.

Chapter 10 Controlling Microbial Growth in the Body: Antimicrobial Drugs

Multiple Choice

1. a
2. d
3. c
4. d
5. b
6. d
7. d
8. c
9. c
10. d
11. a
12. d

Fill in the Blanks

1. *Streptomyces*
2. growing
3. Amphotericin B, fungal cells
4. sulfanilamide
5. narrow-spectrum, broad spectrum
6. can
7. hospitals, nursing homes
8. Cross resistance

Short-Answer Questions for Thought and Review

1. There are so few antiviral agents because viruses are obligate parasites that live inside of host cells and use host cell materials to complete their life cycle. There is very little, therefore, about them that is truly distinct enough to target without killing the host cell outright. Bacteria have many differences from human cells, and therefore present more targets for drugs.
2. Drugs that target general processes such as DNA synthesis are usually not prescribed because they will affect host cells too much (i.e., they will be toxic to the host and the microbe). Drugs that are more specific are a better choice.
3. Clinicians must consider the following when administering drugs to treat infectious diseases: the effectiveness of the drug, the dose needed, the route of administration, the safety of the drug, and the side effects of the drug. Of these, safety and side effects probably weigh most heavily because some patients will be able to tolerate some drugs but not others.

Critical Thinking

1. Of the six categories, drugs that inhibit translation, metabolism, and nucleic acid synthesis, as well as drugs that block host cell recognition, could all be used against viruses. However, the drugs that inhibit translation, metabolism, and nucleic acid synthesis are not always practical because they may inhibit host cell function rather than the virus directly (thus damaging the host while at the same time preventing viral replication). The best drugs to use would be those that target recognition.
2. Antibiotics taken orally must be absorbed through the intestines to circulate in the body. The drug therefore has to be soluble in the presence of *E. coli*. Semisynthetic erythromycin works against Gram-positive and Gram-negative bacteria and can kill *E. coli*. Most oral antibiotics, in fact, list diarrhea as a side effect because of the loss of normal microbiota in the intestine due to absorption of the drug.

3. The MBC test tells you whether the drug you are using is inhibiting growth of or actually killing microbes. Serious infections may require bacteriocidal over bacteriostatic drugs to guarantee success. Bacteriocidal drugs can probably be taken for shorter time periods than bacteriostatic drugs to get the same effect. Knowing what the drug is doing to the microbe in question thus helps clinicians prescribe the right drug, the right dose, and the right time period.
4. Normal microbiota are "adapted" to living in the body spaces they inhabit; they exist well in the physical environmental conditions and the nutrient levels that predominate. Pathogens trying to invade the environment of normal microbiota will not be adapted to the particular situation that exists there and will have to do extra things to get a foothold. Additionally, the physical presence of normal microbiota actually blocks access to host tissues by the pathogens, making it less likely for the pathogens to achieve the initial attachment that is necessary to establishing an infection.
5. Viruses don't have plasmids, pores in their membranes to pump out the drug, or the room to code for enzymes to specifically digest drugs. Viral drug resistance is almost always associated with random mutations in the drug targets that lead to an inability of the drug to interact with the target and prevent viral replication. Because mutations happen at a high frequency in viruses, it is not unreasonable to assume that with constant exposure, multiple "beneficial" mutations will occur that protect the virus from the drug cocktails.

Concept Building Questions

1. In a human host rich in nutrients and energy, it can be hypothesized that the growth curve for a pathogen will show an initial lag stage as the pathogen establishes itself and produces what it needs to grow, followed by a log phase that does not progress to stationary phase. As long as the host exists, nutrients will be prevalent and the pathogen will actively grow. Eventually, with time, an immune response should depress that growth, but it is equally likely that the organism could grow so rapidly that the immune system can't compete. An antibiotic would send the active log phase into stationary phase. Susceptible microbes will die first, leaving the more resistant microbes in the host until the drug reaches a high enough level to kill even the resistant ones. If the person completes drug treatment, it is likely that the drug will kill the pathogen or at least inhibit growth so that the immune system can kill it and the patient recovers (growth curve proceeds to death phase). If a person stops taking the antibiotic too early, then only the susceptible organisms die and the resistant ones remain. As the level of the drug drops, the resistant forms can begin active growth again, reestablishing a log phase of growth (so you would see a "stair step" almost: lag to log to stationary with the drug and then back into log when the drug levels drop).
2. Resistant cells are usually less efficient than normal cells in the absence of an antimicrobial drug because they must expend extra energy to maintain resistance genes and proteins. Therefore, resistant cells are the minority under these conditions because they grow more slowly. When an antimicrobial drug is present, the sensitive cells die (which constitute the majority of cells) and the resistant cells can grow and multiply because they face less competition. Resistant cells then become the majority.

Chapter 11 Characterizing and Classifying Prokaryotes

Multiple Choice

1. b
2. d
3. a
4. d
5. b
6. c
7. c
8. d
9. d
10. c
11. d
12. c
13. c
14. b
15. d

Fill in the Blanks

1. cocci, bacilli, spirals
2. disappears, remaining
3. Crenarchaeota, Euryarchaeota, Korarchaeota
4. Halophiles, Euryarchaeota
5. cyanobacteria
6. *Listeria*
7. metachromatic granules
8. *Streptomyces*, Actinomycetes
9. gammaproteobacteria
10. ammonia, nitrate, *Nitrobacter*
11. chlamydias, negative, peptidoglycan
12. spirochete, Lyme

Matching

1. E
2. O
3. H
4. Q
5. P
6. A
7. F
8. G
9. C
10. D
11. J
12. N
13. K
14. B
15. L
16. M
17. I

Short-Answer Questions for Thought and Review

1. Figure 3.22 summarizes the seven steps in endospore formation. Most of these steps involve assembling new layers of protection around a central core of DNA and cytoplasm.
2. Cyanobacteria are essential because they are responsible for performing a large part of nitrogen fixation, a process that is necessary for turning unusable nitrogen gas into usable nitrate.
3. *Epulopiscium* is quite large relative to other bacteria and thus their size, along with the cilia-like flagella, could have contributed to initial misclassification.
4. Abbreviated descriptions of the five classes of proteobacteria could be written as follows: (1) alpha- and betaproteobacteria differ in rRNA sequences but overlap in metabolic capabilities and the ability to live in low-nutrient environments; (2) gammaproteobacteria live almost everywhere and are the broadest class, being divided into subgroups based on metabolism, oxygen usage, and specializations (purple sulfur bacteria, methane oxidizers, intracellular pathogens); (3) deltaproteobacteria are a small class that is fairly metabolically diverse and includes environmentally important organisms and some pathogens; (4) epsilonproteobacteria also comprise a small group relative to the other proteobacteria and include several examples of pathogenic bacteria.

Critical Thinking

1. Archaea are important to study for many reasons. Two of these reasons are: (1) Archaea are some of the oldest organisms on Earth and by studying them we can learn a great deal about the evolutionary history of our world; and (2) just because they don't cause disease doesn't mean they aren't medically important. They may produce novel chemicals that could, for example, be useful as new antibiotics.
2. In the environment, methanogens produce a lot of methane, but they are generally found in the environment together with methane oxidizers, which utilize much of the methane produced in their metabolism. Methanogens also deposit much of their methane in sediments at the bottoms of the ocean, bogs, etc., and unless these areas are greatly disturbed, the methane remains trapped. Methane production by humans, however, doesn't occur in context with organisms that can clean it up nor is it released into environmental sediments that will trap it. It is released into the air, and so our emissions have a much larger effect on the environment than those of microbes.
3. G + C ratio refers to the number of G and C nucleotides present in a genetic sequence. A low G + C ratio means that there are more A/T nucleotides present; a high G + C ratio means the opposite. Going back and forth would be a big change, but a few mutations here or there are not enough to create the difference between a high G + C and low G + C genome. Thus, long-term divergence led to the different groups we see today. Each "branch" of the tree will be more related to others along its branch, but the two branches would not be expected to be similar because of the sheer number of changes that had to occur over time to cause such a difference in ratios.
4. The gammaproteobacteria, in addition to differences in ribosomal sequences, also differ significantly in terms of metabolic capabilities.

Concept Building Questions

1. Endospores could not metabolize because the thick layers put down to protect the core of the spore are completely impermeable to small molecule traffic. There are no pores, porins, or transporter proteins in the spore coats, nor are the coats diffusible by small molecules. Because nothing goes in, there are no nutrients for catabolism to break down, and without nutrients there are no building blocks for anabolism. Overall, then, there is no metabolism. The impermeability of these coats is also what protects the spore from damage. Lack of transport mechanisms keeps chemicals and other poisons out. Multiple coats, even with weak bonding, present a very tightly bound mass of materials that are not easily disrupted through boiling. Radiation also will have a harder time penetrating the many layers to get to the DNA at the core and damage it.
2. The proteobacteria are the largest and most diverse group of bacteria because they do many different things to earn them that title. Structurally and metabolically they cover just about every imaginable possibility that allows them to survive in all but extreme environments. Genetically, the ability to code for such metabolic and structural diversity is phenomenal (they must have multiple operons dedicated to alternative metabolic pathways, for example).

Chapter 12 Characterizing and Classifying Eukaryotes

Multiple Choice

1. d	5. b	9. b	13. d	17. c
2. b	6. d	10. d	14. b	18. a
3. b	7. a	11. c	15. b	19. c
4. b	8. c	12. d	16. c	20. d

Fill in the Blanks

1. Diatoms, dinoflagellates, dinoflagellates
2. nuclear division, cytoplasmic division
3. diploid, haploid
4. moist
5. trophozoite, cyst
6. foraminifera
7. cell walls
8. mycorrhizae
9. mycelium
10. asexual spores
11. zygomycota, ascomycota, basidiomycota, deuteromycota; deuteromycota
12. phycoerythrin, lower
13. Plasmodial, cellular, myxamoeba
14. helminths, arthropod vectors
15. mosquito

Matching

1. G	4. E	7. C	10. H
2. I	5. F	8. K	11. D
3. J	6. B	9. A	

Short-Answer Questions for Thought and Review

1. The organisms studied as "eukaryotic microbes" are algae, fungi, protozoans, helminths, and vectors (for transmission purposes only). The first four organisms on the list are studied because they are similar to microscopic forms, most have a microscopic stage of life, and many are diagnosed or have infectious forms that are identified microscopically.
2. Meiosis is necessary in sexual reproduction to prevent the number of chromosomes present in an organism from doubling with each generation. Meiosis maintains a stable chromosome number.
3. The key features distinguishing the major groups of protozoa are: (1) alveolates have small membrane-bound cavities called alveoli under their cell surfaces; (2) Rhizaria have threadlike pseudopodia; (3) amoebozoa have lobe-shaped pseudopodia and no shells; (4) euglenoids have plant and animal characteristics and essentially comprise many of the organisms not placed into the other groups; (5) diplomonads lack many internal structures characteristic of eukaryotes; and (6) parabasalids lack mitochondria but have a single nucleus and a parabasal body. Additionally, the groups differ in having, or not having, mitochondria and in the structure of the mitochondria when present. They also differ in where they are found in the environment.

4. Lichens form between fungi and green algae or between fungi and cyanobacteria (most common). The fungus provides protection, nutrients, and water to the photosynthetic symbiont. The algae or cyanobacteria provide photosynthetic products and oxygen to the fungus.

Critical Thinking

1. Protozoans are fairly large microbes and pretty metabolically independent. They don't, in most cases, need to be parasites to survive, and since parasitism is more restrictive than free-living, if you don't have to do it you generally won't. Because they are large, they also run the risk of being seen in the body by the systems designed to protect the body. They may also have a harder time getting into the body in the first place because of their size and structure. Also, the moisture in the human body is not necessarily free for use by the microbes and so may not be much better than the "outside" world.
2. Fungi are different from plants in that they don't have chlorophyll and do not perform photosynthesis; they differ principally from animal cells by having cell walls. Genetically, fungi are more closely related to animals than they are to plants, though for many years they were classified with the plants. Sequence analysis is the best method of classifying fungi as separate from plants and animals, but ultrastructure, morphology, and, to some extent, metabolic capabilities could also be used.
3. *Armillaria* can reach large sizes because of several contributing factors: (1) it lives below the surface of the soil so it is protected from damage; (2) it is saprophytic and feeds off dead organic material, which will be prevalent on the forest floor and in the soil where the fungus is found; and (3) hyphae form massive mycelia and are structured to continually absorb nutrients from the environment and pass them throughout the organism; thus there is good nutrient acquisition and dispersal in the organism.
4. Eukaryotic microbes acquire nutrients by chemoheterotrophism, phagocytosis, absorption (from water or from dead organic matter), and photoautotrophism. There are probably more species among all of the eukaryotic microbes that perform absorption or photoautotrophism than anything else. Photoautotrophs are the basis for all food chains and saprophytic absorption is the basis of recycling. These predominate because they balance each other and utilize two of the most abundant "nutrient" sources on Earth: light + CO_2 and dead organics.

Concept Building Questions

1. The classification of fungi, algae, and protozoa is so complex because of the diversity of structure, which can be confusing within groups and between the different groups; the diversity of metabolism, which puts organisms together in the environment in complex associations that are often hard to decipher; and in terms of genetics, as all are eukaryotic and show many similarities in some key functions. Actual sequencing is really the only useful way to tell many of these eukaryotic microbes apart (DNA, rRNA), with cellular ultrastructure helping in some cases, but not all. Proper classification is important because of the need to be able to treat disease (you need to know whether an organism is a fungus or a protozoan so that the correct drug can be prescribed) and for the ability to understand the ecological roles of the various organisms.

2. Algae perform oxygenic photosynthesis. If the algae were a parasite, it would be inside of another organism in most cases (ectoparasites, those living on the outside, do exist but we will assume an internal parasite here). Inside an organism, it will not be exposed to light. Without light, photosynthesis could not occur. Without photosynthesis, there would be no energy to drive anabolism. Thus, light is critical, and without it the algae cannot survive.

Chapter 13 Characterizing and Classifying Viruses, Viroids, and Prions

Multiple Choice

1. c
2. d
3. b
4. c
5. c
6. c
7. a
8. b
9. c
10. d
11. d
12. c
13. c
14. d

Fill in the Blanks

1. extracellular, capsid, core
2. capsomers or capsomeres
3. helical, icosahedral, complex
4. lysozyme, release
5. lytic, lysogenic
6. poxviruses
7. Parvoviruses
8. retrovirus, +sense RNA, reverse transcriptase, DNA
9. budding, remains alive
10. neoplastic, tumor, metastasis, cancer
11. diploid, continuous

Short-Answer Questions for Thought and Review

1. The steps in the lytic cycle of phage replication are: (1) attachment, where the virus specifically binds to a receptor on the host cell; (2) penetration, where the viral nucleic acid is injected physically into the cell; (3) synthesis, where the virus is replicated, transcribed, and translated; (4) assembly, where pieces of the virus are put together to form functional new virus particles; and (5) release, where virions are released outside the host cell so they can infect new host cells.
2. The key difference between the lytic and lysogenic cycles is that in lysogeny, following attachment and penetration, the viral nucleic acid physically integrates into the host genomic material as a prophage. It remains in the genomic material, repressed and not expressing, until something triggers the end of lysogeny. At this point, the prophage excises itself from the genome and resumes a normal lytic cycle.
3. The differences between phage replication and the replication of an enveloped animal virus are: (1) penetration involves physical injection of DNA in phages and fusion or phagocytosis in animal viruses; (2) animal viruses have to go through an uncoating step to remove the capsid following penetration; (3) synthesis in animal viruses can be in either the nucleus or the cytoplasm, whereas phages always replicate in the cytoplasm (because there is no nucleus); and (4) lytic phages destroy the cell by lysis during release, while enveloped viruses leave the cell by budding.

4. The two major differences between latency and lysogeny are: (1) the prophage always integrates into the host genome but a provirus does not always do so in latency; and (2) if a provirus integrates, it is permanent; it will never excise as a lysogenic prophage will.

Critical Thinking

1. Dichotomous keys are essentially based on physical or metabolic attributes of the organisms being identified. Viruses are similar morphologically and have no metabolism. You could use genetic sequences or replication cycles to create a key, perhaps, but these would not be useful to someone attempting to simply identify a virus for diagnostic purposes.
2. The individual would most likely end up with a viral infection first because the viruses will replicate at a much higher rate than the bacteria will. More viruses means more damage and illness sooner than with the bacteria.
3. If an enveloped virus loses its envelope it will not be able to infect a new cell. The viral spike proteins would be missing (no envelope) and therefore the virus would have nothing by which to attach to the new host cell.
4. From easiest to hardest: dsDNA, ssDNA, +ssRNA, dsRNA, –ssRNA. Minus sense RNA is the genome type that will have the hardest time coming up with mRNA for translation.
5. Viroids and prions have even fewer "life" characteristics than do viruses. Viroids are RNA and nothing else, and prions are proteins and nothing else. Viroids can't replicate on their own at all, can't do anything outside the cell, and appear to code for nothing specific. Prions are mutant proteins that change other proteins and don't carry the genetic change to new hosts. Neither metabolizes, neither can take over a cell, and so they aren't even really good parasites.

Concept Building Questions

1. Viral proteins interact with host receptors to initiate attachment. Very specific interactions lead to very specific host cell requirements and a narrow range of cell types infected. General interactions lead to less specific host cell requirements and a broad range of cell types infected. Specific interactions are mediated by structural interactions between virus and host as well as specific weak bonding. The more bonds formed, the better the structural fit and the fewer "extra" associations that can happen, thus making the fit very specific. Fewer bonds and less precise structural interactions lead to more "extra" associates and a less specific fit.
2. Lysogenic prophages are by default DNA because they can integrate into the bacterial genome. DNA structurally is no different whether one is looking at a DNA virus, a bacterium, or an animal cell. Sequences differ such that each organism makes different proteins, etc., but there is nothing inherent in the sequence to identify viral genes from host genes. Thus, structurally, there is no difference between DNA from different organisms and no way to detect the presence of foreign DNA inside a host cell (except by sequencing everything and trying to sort between them, an almost impossible task). For a lysogenic phage, the only "cure" is to force the virus to excise, hopefully intact, and then capture

the excised pieces and remove them. Such a cure increases the chance of death for the cell, because once excised the virus will quickly initiate replication, which will destroy the cell. Latent animal virus infections can never be cured because there is no way to stimulate excision from the genome.

Chapter 14 Infections, Infectious Diseases, and Epidemiology

Multiple Choice

1. b
2. c
3. d
4. b
5. b
6. d
7. d
8. d
9. c
10. c
11. b
12. c
13. b
14. c
15. a
16. d
17. a

Fill in the Blanks

1. mutualism, commensalism, parasitism
2. axenic, normal flora, resident, transient
3. contamination, infection
4. adhesion
5. pathogenicity, virulence
6. exotoxins, endotoxins, endotoxin
7. zoonoses
8. fomite
9. Biological, mechanical
10. Communicable, noncommunicable
11. descriptive, analytical, experimental

Short-Answer Questions for Thought and Review

1. The factors that contribute to opportunistic infections are: (1) immune suppression; (2) changes to normal flora; and (3) introduction of normal flora into an unusual site in the body. All of these factors have in common the idea that the normal "response" to the microbe has been overturned. Immune suppression prevents active rejection of the invading microbe. Changes to normal flora remove natural "site-specific" antagonism. Introduction into an unusual site "bypasses" both antagonism and potential immune responses.
2. You are most likely to become infected through mucous membranes first, and then through the skin and placenta.
3. *Infection* refers to the invasion (growth) of the pathogen only. *Disease* specifically describes when infection causes enough damage to change normal body function. If the organism does not grow enough to cause damage, then it will never cause disease. Normal flora is an infection, but only under certain extreme conditions does normal flora ever cause disease.
4. The exceptions to Koch's postulates are: (1) some pathogens can't be grown in the lab; (2) some diseases are caused by multiple factors; (3) ethics bars the testing of certain hypotheses in humans; (4) some disease syndromes have more than a single cause; and (5) some pathogens are not well studied.

5. The three general transmission categories are contact transmission (involving close physical interaction of some nature), vehicle transmission (involving an intermediary in transmission between people), and vector transmission (involving animals or insects). Contact transmission probably accounts for most transmission events.
6. Nosocomial infections increase due to exposure (hospitals have high levels of pathogenic organisms circulating), immune suppression (due to illness, surgery, drug treatments, etc.), and accidental transmission between health care workers and patients. Immune suppression leading to opportunistic infections probably accounts for most cases of nosocomial infections.

Critical Thinking

1. Essential mutualistic relationships are probably harder to maintain because not only must the microbe grow in the host, but it must do so while taking minimally from the host and giving back something to the host. There are therefore three restrictions on the organism. With parasitism, the microbe takes and does not give, reducing the restrictions to two.
2. The scenario presents as classic food poisoning, so the trick is to find out who ate what during the day. You would need to catalog the day. Where did you eat lunch and what did you have; did everyone have ice cream and if so all from the same dish/canister of ice cream; what did everyone have at the BBQ? Because five out of eight are sick, five people must have had the same dish at some point during the day and the three who are not sick must *not* have had the same dish. Finding the common source is enough to show causal relationship, as more than 50% of you are sick.
3. This will vary by student. Essentially, though, at some point you would have contacted someone with a cold or something with cold viruses on it (such as a Kleenex). About a week later (after incubation) you would have felt achy, stuffy, and generally unwell (prodromal). You would then go through a variable number of days (3–7) of congestion, watery eyes, itchy eyes/nose, runny nose, maybe a cough (illness). Symptoms would start to get better (decline) and eventually you would be back to normal (convalescence).
4. Based on characteristics, Disease A is the bigger threat. First, there are two reservoirs, meaning that there is a large unknown pool (the sylvatic host) and a close contact reservoir (the cats) from which to get the disease. Second, the mode of transmission is contact and since cats would be in close contact with humans, there is an increased chance of transmission. Though B is transmitted by fleas, unless you are in a situation to be exposed to deer and their fleas, transmission would be less likely than with Disease A.

Concept Building Questions

1. We have seen in past chapters that many viruses are single stranded (thus there is no "check" template) and that there is little proofreading activity during replication. Additionally, viruses replicate more quickly than bacteria and this, combined with poor replicative integrity, produces more mutations per generation than is seen with bacteria. Mutations, however, are random, and so some may be neutral, some may be disadvantageous, and a very few will prove to be an advantage. It is more likely that disadvantageous mutations will crop up. They are

disadvantageous in the sense that point mutations in the genes of spike proteins and capsids lead to slight changes in shape and sequence, which can be enough to distort the ligand/receptor interaction. We have seen previously that even a small change can prevent binding. Most viral progeny, therefore, would be expected to not be able to bind to new host cells. Viruses compensate for this propensity by ensuring the production of huge numbers of progeny.

2. See answers below:
   a. Koch's postulates would seek to isolate the virus from an infected individual and identify it. In this case you could use various diagnostic tests for influenza (can either confirm influenza or, if negative, suggest Disease X). Infection of mice should occur with Disease X but not influenza. Thus you would isolate, identify, and test in an animal. You can't ethically test in humans, even if there is no known mortality associated with an organism.
   b. The portal of entry for the mouse is also most likely inhalation of feces and the portal of exit is feces via the digestive tract. For humans, the portal of entry is inhalation and the most likely portal of exit would be either respiratory or fecal, as in the mouse.
   c. For the first five years of the study, as implied by the characterization of the virus, there is only sporadic transmission of Disease X to humans. Incidence and prevalence are very low because infection is so sporadic. Years 6 and 7, however, show at a minimum a fourfold increase in incidence; this is no longer sporadic and would imply, perhaps, a higher prevalence in mice such that humans are being exposed more frequently (higher incidence in mice and/or larger mouse population). Year 8 shows 220 cases for only half of the year and the assumption would be that the number of cases will continue to rise. The disease is no longer sporadic, incidence is rising and with it prevalence, and at this point we are looking at some type of epidemic. As suggested, increased numbers of mice coupled with increased disease among them could be exposing humans at a higher rate.
   d. The data suggest that there is a shift in the virus/host interaction. The old relationship was between mouse and virus, but it appears that humans are no longer a rare infection step. In terms of mutation, this could mean that the viral ligands have experienced mutations that make infection of human cells preferred over mice (we would have to check to see if the incidence in mice is going down in this case). Such changes would revolve around bonding and structural issues of the ligand. Steps in replication will not have changed (the virus had to be able to replicate in human cells initially and so hasn't suddenly developed that capability) but an increase in initial attachment will allow for more effective colonization of a host.

Chapter 15 Innate Immunity

Multiple Choice

1. d
2. c
3. c
4. d
5. c
6. b
7. b
8. a
9. d
10. d

Fill in the Blanks

1. innate, first
2. second, innate
3. clotting factors, complement, antibodies
4. basophils, eosinophils, diapedesis, emigration
5. Opsonization, opsonins
6. lysozyme, complement, interferon, defensins
7. alpha, beta, gamma
8. pyrogens

Short-Answer Questions for Thought and Review

1. Innate resistance refers to the idea that we are not susceptible to all pathogens because our bodies and physiologies are such that not all pathogens can infect us. For example, we cannot become infected with plant viruses because our bodies are so different from those of plants.
2. The main difference between innate defenses and adaptive defenses is that innate defenses respond to the presence of pathogens in general ways that don't depend on the actual species of invading organism. Adaptive defenses respond only to specific species of organisms.
3. Leukocytes are divided into two groups: granulocytes and agranulocytes. Granulocytes include basophils (which release histamine during inflammation) and eosinophils (which leave the blood to phagocytize microbes). Agranulocytes include lymphocytes and monocytes. Lymphocytes include NK cells (which respond to interferon) and T cells and B cells (both of which are part of adaptive immunity). Monocytes mature into macrophages which are phagocytic. Macrophages can be wandering, alveolar, microglia, or Kupffer cells.
4. Inflammation is caused by the cellular release of histamine, prostaglandins, and leukotrienes. Histamine dilates capillaries and prostaglandins and leukotrienes make them more permeable. Dilation allows more blood flow, which brings macrophages and other cells to the area; increased permeability allows the macrophages to squeeze out of the capillary and enter the tissues.

Critical Thinking

1. Most "surface" infections remain local and non–life threatening because of the regenerative power of the cells infected. Skin and mucous membranes shed constantly, and so even if infected, because the cells naturally detach and are removed at a high rate, they will constantly carry organisms with them and prevent deeper invasion.
2. First-line defenses are essentially physical barriers, whereas second-line defenses involve active cellular or chemical attack. Both are still nonspecific, but the second line is more aggressive toward invading organisms than are the passive first-line defenses, such as skin.
3. Complement causes lysis of either infected cells or microbial cells that have been tagged by antibodies. Interferons are released by virally infected cells and cause surrounding cells to become more resistant to subsequent infection. Defensins actively attack microbial cells by either lysis or inhibition of normal microbial processes. Interferon is the only chemical to work against viruses by creating a nonsupportive environment for replication (it doesn't attack the

virus directly). Defensins work specifically against bacterial mechanisms, none of which are found in viruses. Complement can lyse virally infected cells if the cell has undergone opsonization, but can't attack free viruses.

4. See Table 15.5. Once the outer barrier is breached, local phagocytes would have had to have been unable to trap and kill the bacterium. Eosinophils also would have been ineffective. Inflammation should have drawn cells to the area and promoted a walling off of the site of infection, but either the bacterium made it out of the area (perhaps in the blood) or is again otherwise incapable of being phagocytized. Complement won't work unless antibodies or toxins from the organism are present. If there are no antibodies and the bacterium is Gram-positive and produces no toxin, complement would not have been stimulated. So, once past the skin, if inflammation cannot bring enough phagocytes to the area, or the phagocytes can't engulf the bacterium, it will escape.

Concept Building Questions

1. Bacteria sense attractants and move by biased runs and tumbles toward the attractant (they move to an area of higher concentration). Macrophages must move more specifically because they have to contact the entity they are to engulf. Their sensors are therefore entirely dedicated to "seeing" the cytokines, etc., rather than to responding to any other nutrient-like attractant. They also do not move by runs and tumbles, but will essentially creep along until they physically make contact with their "prey." Both cells must have receptors that internalize signals and translate that into a movement response. Bacteria responding to nutrient attractants can probably get away with using normal transporters on their surface used for the uptake of the same nutrients. Macrophages would have dedicated receptors for the cytokines that are highly specific, such that the macrophage can read the "we need help" signal as well as the signals telling it specifically what to do.
2. See answers below:
   a. If the barrier of the first line of defense in enough hosts is strong enough to repel invasion, then the microbes will not establish infection, will not multiply, and thus will not spread in a population (no increase in microbial number).
   b. Activation of second lines of defense tells the epidemiologist that the agent has a very effective way of making it into the body and past the body surfaces. This will give information about the actual mode by which the agent is transmitted as well as about efficient establishment of infection.
   c. Suppression of symptoms reduces the appearance of disease in an individual and could make that person appear essentially healthy. If there is no obvious outward sign of disease, then an epidemiologist tracking infections would underestimate the number of infected and possibly miss getting to individuals who actually need help.
   d. Even though first and second lines of defense protect us most of the time from getting sick, these lines of defense are highly variable. Not everyone has the same level of epithelial integrity and so the first line will be more easily breached in some. The second line of defense, though active against the invader, is not specific and cannot respond to specific challenges. Individual variability in response will therefore allow the perpetuation of a microbe in a population to a greater extent than would be the case were adaptive immunity not to exist.

Chapter 16 Adaptive Immunity

Multiple Choice

1. b	4. d	7. d	10. a
2. d	5. d	8. d	11. a
3. c	6. c	9. c	12. d

Fill in the Blanks

1. acquired
2. Antigens, epitopes
3. B cells, T cells
4. B, antibodies, humoral
5. answers are as follows
   a. IgG c. IgG, IgM e. IgE
   b. IgE d. IgA f. IgD
6. apoptosis, autoantigens
7. red blood cells, B cells, antigen-presenting cells, red blood cells
8. helper T cells, type 1 helper T cells, cytotoxic T cells, perforin-granzyme cytotoxic pathway, CD95 cytotoxic pathway
9. active, passive, passive
10. active

Short-Answer Questions for Thought and Review

1. The three major types of antigens are: (1) exogenous antigens, which consist of microbial products or structures that appear outside of host cells; (2) endogenous antigens are similar to the microbial products and structures above, but are located inside host cells because that is where the microbe is replicating; and (3) autoantigens, which are normal antigens found on the surfaces of the host tissues and which should not be recognized by the immune system as foreign. Exogenous antigens have the best chance of being seen and so should produce the best immune response.
2. See Figure 16.5 of the book and the accompanying material in the chapter to answer this question.
3. Clonal deletion is absolutely vital to preventing the body from destroying itself. In this process, those lymphocytes that can recognize autoantigens are destroyed. If they were not, the body would continually seek to destroy itself and would have few energy resources left to respond to actual threats. Only with immunosuppressive drugs could an individual survive.
4. Clonal selection is necessary because each B cell is unique and produces unique antibodies. Thus, there may be only one "correct" B cell present. One B cell will produce antibodies, but not nearly enough to clear an infection. Once the right antibody is found it must be mass produced in order to have a hope of controlling all pathogens that might be present. So clonal selection and mass producing the one right antibody are necessary for ultimate clearance of the pathogen from the system.
5. Primary responses are much slower than secondary immune responses. In a primary response, the immune system must find the right cell/antibody to recognize the foreign antigen, must stimulate production of the response element, and then must find and destroy all antigens. The initial finding of

the right immune cells for the job takes some time because there is no past history to rely on. Secondary responses are memory responses. The antigen has already been seen, the immune system has already been told to watch for it, and so as soon as the antigen is seen again, memory cells automatically proliferate and contain the pathogen, usually fast enough that no illness occurs.

Critical Thinking

1. The immune system does not actively patrol the brain or spinal cord because these systems are essentially separate from the tissues and cells of the immune system. Lymphatic capillaries are not found in the brain or spinal cord, so no lymph is delivered to these areas (no lymphocytes), nor are there immune tissues actually present in these areas. Antibodies and cells don't pass the blood-brain barrier either, so these sites are devoid of immune intervention.
2. Without B cells you would have no antibodies and so you couldn't do any sort of agglutination, neutralization, or opsonization activities that relied on antibodies. You would have no B cell memory, either. You would also be lacking in some forms of complement stimulation and would not have classic allergic responses.
3. The bacterium is first phagocytosed by a macrophage that processes the antigens and places them on the surface in conjunction with MHC II. The macrophage then presents to the appropriate type 1 helper T cell, which causes it to become a type 2 helper T cell. Helper cells stimulate B cells to divide and produce antibodies and stimulate cytotoxic T cells to kill the cell showing the antigen.

Concept Building Questions

1. DNA and other internal structures of cells and viruses are "foreign" in the sense that, outside of their microbial "vessel," they are out of place. DNA is not supposed to be free in the body. There is nothing on DNA to mask it from the immune system, identify it as self, or otherwise hide it. It will be seen as being out of place and will be responded to vigorously. The same would be true for organelles and proteins normally found inside the microbe. However, though all of these internal structures would elicit a strong response, they would not be protective because in the normal course of infection, these structures are never seen. They are not on the outside, and so to have antibodies present against DNA will not neutralize or otherwise stop the infection because only after the microbe dies would the antibodies be able to see the DNA inside. Thus, something can be immunogenic, but not offer protection.
2. Though derived from external structures, the antigen/epitope is much more specific because of the relationship it has with antibodies and other components of the immune system. It is the specific interaction, based on bonding and conformation, that presents a much more precise method of identification and classification. Only one or a very few antibodies will recognize and bind to a given antigen. Though the protein may be common to bacteria, different bacterial species will have slightly different versions of the protein (thus different determinants) and will elicit slightly different antibodies. They can therefore be distinguished based on antigen/antibody bonding where they couldn't be distinguished based on the protein alone.

Chapter 17 Immunization and Immune Testing

Multiple Choice

1. b	4. c	7. a	10. d
2. d	5. c	8. c	11. c
3. b	6. c	9. c	12. a

Fill in the Blanks

1. attenuated, inactivated, toxoid
2. attenuated
3. Inactivated, Adjuvants
4. Antiserum, antitoxin
5. immunodiffusion, immunoelectrophoresis
6. insoluble, soluble
7. Hemagglutination
8. lyse, antibodies, antigens, anti-RBC antibodies
9. antigen, antigens, antibodies
10. immunochromatographic

Short-Answer Questions for Thought and Review

1. There are many answers for this question, only a few of which are given here. Politically, it is difficult to convince people to take money from one source to pay for vaccines to be used outside of their own jurisdiction (i.e., first world countries paying for the third world). Along these lines, the private companies that make the vaccines will only make those that are profitable, thus basing much of their research and development on purely economic grounds. Socially and culturally, some societies/peoples don't readily accept the idea of vaccination, making administration difficult even if the vaccine is available. Scientifically, it is difficult to make vaccines against many microbes because they change too much. All of these either reduce the number of available vaccines or interfere with distribution, thus the constant presence of disease among us.
2. Modern vaccine technology relies on molecular biology and the ability to create recombinant molecules (molecular biology is the foundation of all new techniques). Some ideas include the selective deletion of virulence genes, production of recombinant subunit vaccines, production of multivalent vaccines through the recombination of microbial genes, and DNA vaccines. All of these, in addition to relying on molecular biology, seek to reduce the microbe to a few parts that would not be infectious.
3. Immunoelectrophoresis gives better sensitivity and specificity because it separates out antigens better and can be quantified more readily (the precipitate lines) because of the separation capabilities of electrophoresis. Immunodiffusion really relies on diffusion and is thus less reproducible and harder to quantify (or, in some cases, qualify).

Critical Thinking

1. The second cold you get may be milder than the first, especially if it follows fairly soon after the first, because of partial immunity. There are many cold viruses out there, but a great many of them are related. So you won't be

protected from the newest version of the cold, but the antibodies/immune cells left over from the first episode have a higher probability of being cross-reactive than not. This can lessen the symptoms of the second cold by restricting replication of the virus.

2. Basically you would need to design some sort of immunoprecipitation/ immunoelectrophoresis or other immune test that would allow you to (1) detect the presence of antibodies against the vaccine agent (showing that the vaccine worked in some fashion) and then (2) determine the level of antibodies present (the higher the level, even after the initial vaccination period, the more likely that the stimulus was enough to engender strong immunity). You can't test the vaccine by challenge (exposure), but you can gauge response by comparing the above to other vaccines with known responses and successes.
3. For Figure 17.3, use Table 17.2 to define bacteria vs. virus and the type of vaccine used. Any vaccine that is not attenuated will require boosters, though you could argue that only those microbes you are most likely to encounter should be the ones you get boosters for if you want to minimize repeat shots. Tetanus, for example, should routinely be given every 10 years because the probability of injury and contamination of the wound is so great. If you don't remember when you last got vaccinated against some of these agents, it may be time to get revaccinated.
4. You need at least two, if not more, antibody time points to determine if someone is just beginning an infection (points go up), is at peak infection (points rising slightly but most likely leveling off), or is recovering (points declining). Someone might naturally have high or low levels of certain antibodies, so the level alone is not diagnostic. Comparing two or more values, taken over the course of a few days, will tell you which direction your titer is going.
5. ELISAs rely on detecting some reaction (fluorescence or chemical reaction) over background. Early in the infection, though antigen may be present, it may not be present at high enough levels to get over the background of the assay system. Later in infection, or even after infection, if you are checking for antibodies, you will get noticeable (statistical) differences from background, and thus a more reliable result.

Concept Building Questions

1. Some possible answers are as follows:
   a. Split diagnostic confirmation and antibody confirmation into two tests. In the first test use known Bacterium X in one well and a patient sample in another; the precipitate line indicates antibodies in serum to a known antigen. In the second test, use anti-antibodies. Place patient serum in one well and surround with wells for the five anti-antibody classes; precipitate lines indicate which antibodies are present. In both cases, known antigens and antibodies need to be used as controls or references.
   b. For respiratory exposure, you would expect IgA because this is the predominant antibody in the area. For the blood, primary exposure would elicit IgM.
   c. Recognition is based on noncovalent bonding and conformational matching. Full recognition shows tight and specific binding; partial

recognition will show less specific binding. You could, as one possible example, design a complement inhibition test to tell the difference. Full recognition should be better able to inhibit complement-mediated lysis of RBCs than partial recognition, particularly over varying dilutions of antibody.

Chapter 18 AIDS and Other Immune Disorders

Multiple Choice

1. a
2. a
3. d
4. d
5. a
6. d
7. a
8. d
9. b
10. d
11. d
12. c
13. c
14. d
15. a

Fill in the Blanks

1. hypersensitivity, immunodeficiency
2. mast cells, basophils, eosinophils
3. hemolytic anemia
4. autografts, isografts, allografts, xenografts, xenografts
5. T, single-organ, systemic
6. multiple sclerosis
7. Opportunistic infections
8. genetic, developmental, malnutrition, severe stress, infectious disease
9. HIV-1

Short-Answer Questions for Thought and Review

1. Degranulation occurs when an allergen binds to a sensitized cell that has surface IgE. Binding triggers a chemical cascade that stimulates the release of inflammatory mediators from granules lying just under the plasma membrane. The inflammatory mediators cause swelling of mucous membranes and other irritations (to draw immune cells to the area), leading to itchiness, watery eyes, runny nose, etc.
2. In type II hypersensitivities, complement is activated by antibodies that bind to drug molecules bound to the surfaces of platelets, leukocytes, or red blood cells. In type IV hypersensitivities, small molecules bind to any protein they contact on the surface of cells (particularly skin cells), triggering cytotoxic T cell–mediated destruction of the small molecule-bound cell.
3. Autoimmune disease may occur more frequently in the elderly because as we age, our immune systems begin to fail. The system doesn't respond as well, as strongly, or as readily to shut-down command. Clonal deletion processes will fail within this environment as time goes by, and so cells that should be deleted can slip through.

Critical Thinking

1. Histamine is only one of several inflammatory mediators released in response to hypersensitivity reactions. The antihistamine only takes care of histamine, leaving the other mediators to continue to stimulate inflammation, pain, and

redness. Kinins, proteases, prostaglandins, and leukotrienes are all produced by mast cells and will continue to be released even if the histamine is neutralized. They stimulate more degranulation, rapidly outpacing the effect of any topical agent. Basophils and eosinophils will also continue to stimulate inflammatory responses.

2. Type AB individuals have both A and B antigens; type O individuals have no antigens. Thus, AB individuals could experience problems with cross-reaction due to exposure to certain microbes. The ultimate danger of such infections is that if enough microbes mimic something in the body, our immune system will rapidly lose the ability to distinguish true threats from benign antigens and won't be able to protect us as it should.
3. The major disadvantage of immunosuppressive drugs is that they depress your ability to fight off infections over the long term. The more the body is prevented from fighting off infections, the greater the chance that something serious will be encountered. We encounter microbes constantly, and only the routine suppression of these microbes by active immunity keeps us from being sick all of the time. For someone on immunosuppressive therapy, routine patrol is depressed, leading to insufficient prevention of microbial colonization.
4. Yes, an autoimmune disease can be an immunodeficiency disease. Attacking oneself can lead to destruction of immune cells needed to fight off infections. By removing these cells, or the cells that give rise to them, there isn't anything left to stimulate defenses against invaders. Thus, you could mount a deficient immune response to foreign pathogens while at the same time actively destroying your own cells.

Concept Building Questions

1. Immunotherapies work to desensitize the immune system to responding to benign antigens but they do so at the protein level, rather than the genetic level. Immunotherapies don't do anything to change the nature of the immune cells produced per se (does not alter genetics, clonal deletion or clonal selection processes, etc.). They merely show an antigen to the immune system to the point that the system gives up responding to it. Over time, as the level of introduced antigen drops, the immune system rebounds to where it was before the antigen came; now each time it sees the pet dander, for example, it resumes reacting to it.
2. Privileged sites are a benefit to graft transplants because the graft goes into a place that is not patrolled by the immune system, thus there can be no rejection of the foreign tissues. Unfortunately, there can also be no active patrolling for pathogens, either. The immune system does penetrate to places like the brain and so immune cells can't get there to deal with problems. Whereas this protects the graft, it also protects the microbe, particularly viruses that are small enough to easily bypass the blood-brain barrier to reach the brain tissue itself. The reason there aren't a lot more infections in the brain, nervous system, or other privileged sites is because exit from the host from these sites is difficult, which hinders the survival capabilities of pathogens.
3. Viruses are intracellular parasites. They take over cells and make cells do their bidding. They insert their proteins into the cell, can alter the cell's DNA, and generally threaten to change the infected cell to the point that it is functionally different from what it should be. In many cases this difference is seen and then destroyed by the immune system, but depending on what the

virus does, it could alter the surface of the cell enough to trigger the immune system not into responding against the virus, but into responding against the cells themselves. This can lead to "extra" damage and autoimmune disease.

Chapter 19 Microbial Diseases of the Skin and Wounds

Multiple Choice

1. c
2. d
3. d
4. b
5. d
6. a
7. a
8. d
9. c
10. d
11. a
12. c
13. b
14. c
15. d
16. b
17. d
18. a

Fill in the Blanks

1. microbiota
2. melanin
3. epidermis
4. Coagulase
5. vancomycin
6. *S. aureus* and *S. pyogenes*
7. superficial, cutaneous, subcutaneous, systemic
8. endotoxin
9. herpes simplex virus type 2
10. chickenpox, shingles
11. respiratory droplets, fluid from lesions
12. erythema infectiosum
13. herpangina
14. measles (rubeola)

Matching

1. B
2. I
3. D
4. G
5. F
6. E
7. H
8. C
9. J
10. A

Short-Answer Questions for Thought and Review

1. The skin is inhospitable for a few reasons. It is covered with salt and sebum. Sweat and sebum contain compounds that are antimicrobial. The outer skin layer has dead, flat, dry cells that are a barrier to microbial invasion. Additionally, the outer layer of skin is continually replaced and thus microbes attached to this layer of skin are removed as well.
2. The capsule present on some bacteria is capable of inhibiting chemotaxis as well as preventing phagocytosis of the bacteria by leukocytes. This slime layer is also useful in helping bacteria attach to surfaces (such as medical devices) where they can become a source of infection.
3. *Pseudomonas aeruginosa* is commonly found in the environment. However, it is not often a cause of disease because intact skin and its chemical defenses are usually impenetrable to the bacterium. This is why *P. aeruginosa* is a common infection in burn patients who have had the normal barriers provided by the skin compromised.

4. Certain enzymes produced by strains of *S. pyogenes* allow the bacterium to invade body tissue. These enzymes include deoxyribonucleases, streptokinases, and hyaluronidase. Toxins such as exotoxin A and streptolysin S are also produced that aid the virulence of the bacteria.

Critical Thinking

1. Viruses must enter their hosts' cells in order to produce progeny. Entry into the cell is facilitated by receptors on the cell surface. The correct receptors must be present in order for a virus to attach to a particular cell and gain entry. Cowpox and monkeypox are not as capable of attaching to the receptors on human cells and thus they are less likely to gain access to the human body and cause disease.
2. Further exposure to herpes simplex virus was not needed to produce the lesions. The lesions have recurred because the virus was not eliminated from the body during treatment with medication. The herpesviruses are capable of remaining latent in the human body and thus treatment does not cure an individual of the virus. Periodically the virus can reactivate and again produce symptomatic disease.
3. Although chickenpox and shingles are caused by the same virus (varicella-zoster virus), an individual cannot catch shingles from someone with chickenpox. This is because shingles results from a reactivation event of a latent virus that is already in the body due to a previous exposure to VZV.

Concept Building Questions

1. Nails actually grow from "the inside"; the part you see is dead, but the part that initiates it is not. Topical treatment of fungi on the dead part of the nail will not kill the actual source of the infection, which is in the nail bed where growth occurs. Thus, you need to treat orally so that the antifungal agent can build up in the nail (accumulate as it grows) and kill the fungus from the inside out.
2. There are many possible answers for this question; only two have been given here. One reason to destroy the known stocks is to prevent accidental release of the virus back into nature (accidental inoculation in a laboratory setting or accidental release due to political instability that breaks the chain of control of the agent). One reason to keep the stocks is for research into better vaccines and the development of treatment options. This reason looks to the future, when another poxvirus like smallpox could arise due to mutations in other poxviruses.

Chapter 20 Microbial Diseases of the Nervous System and Eyes

Multiple Choice

1. a
2. d
3. b
4. c
5. d
6. c
7. b
8. a
9. d
10. a
11. b
12. d
13. c
14. b
15. d
16. c
17. a
18. b

Fill in the Blanks

1. neuroglia, neurons
2. axenic
3. tuberculoid leprosy
4. Humans, nine-banded armadillos
5. food-borne, infant, and wound botulism
6. acetylcholine
7. *Enterovirus*
8. flaccid paralysis, contract
9. *Chlamydia trachomatis*
10. arthropod-borne virus
11. tsetse fly
12. cornea
13. Trachoma
14. pinkeye

Matching

1. D	3. H	5. C	7. B
2. F	4. G	6. A	8. E

Short-Answer Questions for Thought and Review

1. Signals travel down the axon to the terminal end of the axon. Here junctions called synapses are formed with other cells. The signal is released in the form of neurotransmitters to the postsynaptic cell, usually across a synaptic cleft. This either stimulates or inhibits some action.
2. Microbiota can gain access to the central nervous system a number of ways. Breaks in the bones or meninges as well as medical procedures can create points of entry for pathogens. Pathogens in the blood or lymph can infect and kill the meninges. Inflammation can alter the permeability of the blood-brain barrier, providing access for microorganisms.
3. Infant botulism is caused by the same bacteria and toxins; however, the mode of infection is different. Food-borne botulism results from ingestion of preformed toxin in contaminated food. This toxin produces the symptoms characteristic of the illness. Infants acquire botulism by ingesting the bacteria themselves rather than the toxin. The bacteria are able to grow in the intestine and then secrete the toxin that causes illness.
4. Poliovirus can manifest in four ways. Most poliovirus infections are asymptomatic and thus don't produce signs of illness. Minor polio produces symptoms but they are temporary and nonspecific. Nonparalytic polio produces symptoms like minor polio but also causes muscle spasms and back pain. Paralytic polio is the most severe and causes varying degrees of paralysis. The paralysis will resolve in time for most cases although some patients have permanent paralysis.
5. A prion is an infectious protein. An abnormally folded prion can act to "catalyze" the transition of a normal prion protein into an abnormal prion protein. A misfolded prion uses surrounding, normally folded, prions to make new abnormal ones. This process results in destruction of the brain.

Critical Thinking

1. Intoxications are often more deadly than infections because of several factors: (1) toxins are produced in excess of the cell and are either secreted or released upon microbial cell death; this means that they reach high levels fairly readily; (2) toxins are often harder to destroy than vegetative cells, whether you are cooking something or trying to treat someone who has ingested the toxins; this means that they have a better chance of binding to cells, altering functions, and causing damage than of being prevented from doing so; and (3) there is less opportunity to survive an initial exposure to toxins damage-free so as to achieve immunity to further exposures.
2. The Salk vaccine is a dead vaccine that cannot be transmitted from the vaccinated individual to others or be shed into the environment. In eradication, you need to eliminate natural reservoirs of the virus in the environment. Using a live virus vaccine, such as the Sabin vaccine, could continue to propagate the virus between humans, allow for recombination of the vaccine strain and the wild-type strain of the virus in the environment, and thereby keep the virus perpetuating among humans. With the Salk vaccine there is no shedding and thus no possibility of having the virus reenter the population.

Concept Building Questions

1. The five bacteria commonly causing bacterial meningitis all have some form of virulence factor to aid their ability to cause disease. A common virulence factor is the presence of a polysaccharide capsule. Having such a capsule allows the bacteria to avoid phagocytosis and digestion. Individual species produce additional virulence factors. *Streptococcus pneumoniae* strains can have proteases that destroy IgA. *Neisseria meningitidis* has fimbriae and lipooligosaccharide that allow the bacteria to attach to cells. *Listeria* produces listeriolysin O that enables the bacteria to escape the phagosome and avoid digestion.
2. A bacterium that has been phagocytized is usually digested by the cell. However, *Listeria monocytogenes* produces the enzyme listeriolysin O that allows it to escape the phagosome before being digested. The bacteria is then able to move to the cell surface where it becomes enclosed in a pseudopod that is ingested by a neighboring cell. This allows the bacterium to move from cell to cell without being detected. *Trypanosoma brucei* is also able to evade the immune system, but in a different manner. The protozoan changes the glycoproteins in its surface. These act as antigens for recognition by antibodies. However, in this case, by the time antibodies have been made against a particular glycoprotein, new glycoproteins have been produced and thus the antibodies cannot recognize the new antigens. This cycle continues and the immune system is never able to "catch up" and recognize the ever-changing glycoproteins.

Chapter 21 Microbial Cardiovascular and Systemic Diseases

Multiple Choice

1. b
2. c
3. d
4. a
5. d
6. b
7. a
8. c
9. a
10. d
11. b
12. c
13. a
14. d
15. c
16. b
17. b
18. a

Fill in the Blanks

1. cardiovascular system
2. osteomyelitis
3. Septic shock
4. nosocomial
5. occult septicemia
6. b-lactamase
7. buboes
8. Lyme disease
9. Leukopenia
10. Africa
11. Jungle monkeys
12. Schizogony
13. Cats
14. Pseudocysts

Matching

1. C	3. G	5. A	7. D	9. E
2. H	4. J	6. I	8. B	10. F

Short-Answer Questions for Thought and Review

1. There are various genetic traits that contribute to an individual's increased resistance to malaria. Sickle-cell trait, in which the erythrocytes are sickle shaped, prevents the cells from being penetrated by *Plasmodium*. Individuals with two genes for hemoglobin C are protected by unknown mechanisms from malaria. Individuals deficient in glucose-6-phosphate dehydrogenase don't support DNA replication of trophozoites. Finally, persons without Duffy antigens on erythrocytes can't be infected because *P. vivax* requires these antigens on the cell surface to infect them.
2. *Borrelia* species use manganese in their metabolic processes rather than iron. By not using iron at all they circumvent the problem of limited iron availability. This is significant because for almost every other microbe, iron limitations limit growth. In the host, *Borrelia* doesn't have to compete for iron and can grow without the limitation, thus more easily reaching levels dangerous to the host.
3. Cercariae present in water penetrate human skin. Adults develop and mate in blood vessels of the host. Eggs are then passed from feces or urine into the environment, where the eggs hatch into miracidia in aqueous environments. These miracidia infect snails, where asexual reproduction occurs, producing the cercariae that once again can infect a human host.

Critical Thinking

1. In theory, at any point during the *Plasmodium* life cycle where the parasite is extracellular, it could be recognized by antibodies and neutralized. The times this would occur would be when the parasite travels to the liver cells initially, then en route from the liver to the RBCs, and also when traveling between RBCs. Giving someone antibodies before being bitten or immediately after being bitten may prevent

establishment of the parasite in liver cells. Giving someone antibodies during episodes could control the infection probably to a good extent because the cycles of cell lysis are synchronous. Assuming no dormancy in the parasite, you might be able to neutralize and clear all free parasites to stop the infection.

2. Although dengue fever can be associated with extreme pain it is usually a self-limiting disease. Dengue hemorrhagic fever, on the other hand, can be associated with a much worse outcome. This is because dengue hemorrhagic fever is a hyperimmune response that occurs when someone becomes infected a second time with the same dengue virus. When this occurs, the antibodies that were produced in response to the first infection can form complexes with the newly infecting virus. These complexes are phagocytized cells that activate memory T cells. These T cells then release a large amount of inflammatory cytokines that can produce hemorrhaging and shock and can lead to death.
3. Many of the eggs look very similar and only those with extensive experience will be able to consistently identify helminths based on this. Because helminthic infections are rarely fatal, are relatively rare in the United States, and their signs and symptoms are often diagnostically helpful, there is little push for more advanced and expensive diagnostic tests.

Concept Building Questions

1. With mono, the virus infects B cells; T cells then try to kill the infected B cells and this gives rise to the "civil war" seen in the immune system in response to the presence of the virus. The better the immune system is at recognizing and attacking infected cells, the more symptomatic the individual. Therefore, someone who is immunocompromised should display fewer symptoms because that person's immune function is lower overall. Hypersensitivities imply overactive immune function. Depending on the type of hypersensitivity the individual normally manifests, symptoms could be aggravated.
2. The infectiousness of plague increases with amplification in nonhuman reservoirs and insects/ticks/fleas. Amplification cycles essentially provide an optimum host in which the microbe can grow to large numbers. These amplification reservoirs are usually numerous and rarely become ill and so continue to propagate the microbe. An increased presence of bacteria in the environment in other animals or insects increases the chances that someone may be exposed. People are therefore more at risk any time a disease uses animals or insects as an amplification cycle.

Chapter 22 Microbial Diseases of the Respiratory System

Multiple Choice

1. c
2. d
3. d
4. a
5. b
6. c
7. a
8. c
9. d
10. b
11. c
12. b
13. d
14. d
15. a
16. b
17. a
18. c

Fill in the Blanks

1. nose, nasal cavity, pharynx
2. Streptokinase
3. oxygen, carbon dioxide
4. otitis media
5. Pneumonia
6. Mycoplasmal (primary atypical) pneumonia
7. tuberculin skin test
8. Hemagglutinin
9. syncytia
10. metapneumovirus
11. amphotericin B
12. *Pneumocystis* pneumonia
13. paroxysmal

Matching

1. G	3. F	5. A	7. C
2. H	4. B	6. D	8. E

Short-Answer Questions for Thought and Review

1. *Streptococcus pneumoniae* hide from the immune system by producing a protein on their cell wall called phosphorylcholine that binds to receptors on certain cells (lungs, meninges, blood vessel walls). This causes those cells to phagocytize the bacterial cell. By forcing nonphagocytic cells to take them in, they remove themselves from visible patrol by the immune system.
2. One polypeptide of the diphtheria toxin binds to a human cell and results in endocytosis of the bacteria. The toxin is cleaved, activating the second polypeptide. This polypeptide destroys the protein, elongation factor, which is needed for polypeptide synthesis in eukaryotic organisms. This toxin is very potent because a single toxin molecule destroys multiple molecules of elongation factor. When this occurs, polypeptide synthesis in the host is completely shut down and the cell dies.
3. Colds and influenza can seem similar but there are differences in the symptoms. Common features of both a cold and influenza include congestion, malaise, and a cough. However, influenza is often differentiated from a cold by the presence of a fever. Additionally, influenza can cause myalgia, or muscle pain, that is not experienced by an individual with a common cold.

Critical Thinking

1. Rhinoviruses are limited to the upper respiratory tract because they only replicate at low temperatures and thus can't go anywhere else in the body that is a warmer temperature. This limits the severity of infection because once the outer layer of cells is gone, there is nowhere else for the virus to replicate, thus making such infections self-limiting.
2. Tuberculosis fights a "war" with the immune system. Upon infection, the microbes begin to grow in the lungs and the macrophages and other immune system players begin to fight. If the immune system fails, the

microbe invades further. If the immune system wins, the microbe is cleared. In an immunosuppressed individual, initial infection won't be fought off and there is more of a chance the infection will become severe and spread. Loss of the suppressive function of the immune system to keep the microbe in check as someone becomes immunosuppressed or immunocompromised means that the microbe has a better chance of breaking out of its "prison" to grow and spread.

3. Antigenic drift and antigenic shift are both processes that result in mutations in the surface glycoproteins, hemagglutinin (HA) and neuraminidase (NA). Antigenic drift indicates the accumulation of mutations in HA and NA within a single strain of virus in a particular geographic region. This occurs at a more rapid pace than antigenic shift. Antigenic shift produces significant antigenic change due to the genome reassortment among different influenzavirus strains within the host cells. Although the changes in this case are more significant, these events occur at a considerably slower rate than antigenic drift. Influenzavirus type B would undergo antigenic changes by antigenic drift but not antigenic shift. This is because influenza B occurs primarily in humans and thus major alterations in the genome of viruses infecting different hosts would not occur.

Concept Building Questions

1. *Legionella pneumophila* is a bacterium that is ubiquitous in water environments, including places such as air-conditioning ducts. This is somewhat unusual, given the nutrient requirements needed by the microorganism for survival. *Legionella* handles this problem by existing as an intracellular parasite when in the environment. The microorganism infects and reproduces within protozoa, such as amoebae. The bacteria are present within vesicles and can infect humans when these vesicles are inhaled. *Legionella* is further able to survive in aquatic environments because it is resistant to the heat and chlorination used to eliminate microorganisms in these locations.
2. A positive tuberculin test tells you that the individual has come in contact with tuberculosis, but the test alone cannot tell you anything more than that. The test is a type IV hypersensitivity reaction and involves memory T cells that respond to second exposures to an antigen. Tuberculin is a protein solution derived from *M. tuberculosis*; it mimics the structure the immune system would normally see but without the danger of infection. The presence of memory T cells could be the result of an old primary exposure or they could be the result of the presence of active infection. There is no way to tell the difference except to use other tests to confirm the presence of active tuberculosis infection.

Chapter 23 Microbial Diseases of the Digestive System

Multiple Choice

1. d
2. b
3. c
4. a
5. b
6. c
7. b
8. d
9. a
10. d
11. a
12. b
13. c
14. a
15. b
16. c
17. d
18. a

Fill in the Blanks

1. peritoneum
2. Peristalsis
3. viridans streptococci
4. microbial antagonists
5. fluoridation
6. urease
7. bacterial gastroenteritis
8. Enterotoxins
9. cold sore
10. trigeminal nerve ganglia
11. hepatitis
12. Sexual contact
13. luminal amebiasis
14. tapeworm

Matching

1. F	3. G	5. J	7. D	9. I
2. H	4. B	6. C	8. A	10. E

Short-Answer Questions for Thought and Review

1. Cholera toxin is an AB toxin; the B subunits bind to receptors on the intestinal epithelial cells and release the A subunit into the cell. The toxin acts as an adenylate cyclase to convert ATP to cAMP, a signaling molecule that stimulates the cells to release electrolytes at a high rate. As the electrolytes leave the cell, they draw water out as well. This ultimately leads to severe dehydration, acidosis, and shock. These symptoms can cause death if not treated.
2. HBV protects itself from immune clearance by secreting viral antigens that act as decoys. Spherical and filamentous particles are capable of binding antibody, which lessens the antibody available to bind to Dane particles, which are the actual infectious virions.
3. Food poisoning due to intoxication such as that caused by *Staphylococcus aureus* has a rapid onset and is cleared just as quickly. This is explained by the idea of intoxication. The cause of illness is the toxin that is ingested in contaminated food. Because the toxin is harmful upon ingestion, symptoms result very rapidly. Microbes often must replicate in the host before symptoms develop, thus disease onset is delayed. Intoxications are quickly resolved because as soon as the toxin is eliminated from the body, the trigger causing symptoms is gone and thus so are the symptoms.

Critical Thinking

1. In the case of self-limiting infection, it is generally best not to prescribe antibiotics for two essential reasons: (1) overuse of antibiotics can lead to microbial resistance and this resistance can be transmitted to other microbes and (2) antibiotics will kill normal flora as well, which in turn decreases antagonism and thus increases the chances that the pathogen will find the ability to grow.
2. *Giardia* can persist for months in the environment outside of animal hosts and is carried by more than beavers. Just because no beavers are in sight

doesn't mean that *Giardia* isn't still present. Drinking upstream is not safe because other animals could still be in the area depositing *Giardia* into the water.

3. The child who lives in the country could potentially have more cavities. This is because the water she drinks is likely not fluoridated. Fluoride present in fluoridated water is incorporated into the enamel of developing teeth and protects them from the effects of acid. Therefore, the child consuming water without fluoride has teeth that are more susceptible to the effects that cause tooth decay and result in dental caries.

Concept Building Questions

1. Since these parasites are often transmitted by the fecal-oral route or by consumption of contaminated food or water, they are traveling through the harsh environment of the stomach. The cyst is a resistant form and stands a much better chance of surviving the trip than the cellular trophozoite, which lacks the protection of the cyst structure.
2. Acquiring bacterial gastroenteritis is often the result of fecal-oral transmission of the pathogen. Thus, individuals become sick due to the consumption of fecally contaminated water or food. This is more likely to occur in developing countries. These countries often have poor sanitation practices, and people may not have access to clean water. Additionally, people may live in close contact with animals that also might transmit some of the bacteria responsible for this disease.

Chapter 24 Microbial Diseases of the Urinary and Reproductive Systems

Multiple Choice

1. b
2. a
3. d
4. b
5. c
6. a
7. c
8. d
9. d
10. b
11. d
12. a
13. c
14. b
15. d
16. d
17. a
18. c

Fill in the Blanks

1. uterus
2. Dysuria
3. enteric bacteria
4. glomerulonephritis
5. MHC class II, T cells
6. pseudohyphae
7. uterus
8. syphilis
9. latent syphilis
10. Chlamydia
11. soft chancres
12. Condylomata acuminata
13. trichomoniasis

Matching

1. F	3. A	5. E	7. D
2. H	4. G	6. B	8. C

Short-Answer Questions for Thought and Review

1. Women tend to get urinary tract infections more often than men due to anatomical differences. Women have shorter urethras than men have. Additionally, the urethra of a female is closer to the anus, which is often the source of bacteria causing urinary tract infections.
2. Condom use can often help prevent the spread of some sexually transmitted diseases. Because herpes lesions in women are usually on the external genitalia, use of a condom during sexual intercourse provides little protection to their partners. People with active lesions should abstain from sexual activity until the crusts have disappeared.
3. Toxic shock syndrome toxins (TSSTs) are responsible for the symptoms seen with toxic shock syndrome. The toxins bind to major histocompatibility complex II molecules and T cells. This results in T cell activation. Due to the large number of toxin molecules, more T cells are activated than would occur in a normal immune response, and these cells produce a large amount of cytokines that trigger the signs and symptoms of the disease.

Critical Thinking

1. This person cannot be said to not have syphilis. This is because the methods used to detect the bacterial cause were faulty. The cause of syphilis is *Treponema pallidum*. First, this bacterium has never been cultured on an agar plate. It can only be grown in cell culture. Additionally, *Treponema* is a very sensitive and fastidious organism. *Treponema* cannot survive in the environment for long periods and thus would not have survived the trip to a laboratory for detection.
2. If a mother does not have an active herpes lesion at the time of delivery it generally means that she is in a latent phase of the disease. If latent, the virus is quiet in nerve cells and would not come in contact with the baby as it is being delivered. However, a cesarean section would still be advisable, even with treatment, to completely avoid the risk of any virus contacting the baby.
3. *Chlamydia* has a unique life cycle because the bacterium exists in two different forms at different stages in the life cycle. Elementary bodies (EBs) are the infective form of the bacterium and thus exist outside the body and can survive environmental conditions well. The reticulate bodies (RBs) are the reproductive form of the bacterium that exists within the body and divides to produce additional bacterial cells. EBs enter the host via phagocytosis. Once inside the cell, the EBs can convert to RBs inside the phagosome. The RBs then divide to produce more RBs. The RBs then convert back to EBs, which are released and available to infect new cells.

Concept Building Questions

1. The lack of symptoms is beneficial to the bacteria because it can continue to spread from unknowing hosts as it continues to grow inside of them. The lack of symptoms will not cause any change in behavior of the host so as to limit spread, there won't be any treatment, and the bacteria are thus allowed

a longer period in which to grow and divide. This ultimately increases the chances of survival and movement to the next host. This is a huge disadvantage to health care workers, however. If hosts do not realize they are ill, they will not seek treatment or limit their activities. They will continue to spread the disease through the population and by the time the health care worker becomes aware of the problem, it is a large problem instead of a small one.

2. This treatment will not be valuable. Antibiotic treatment is useful in all other stages of syphilis except the tertiary stage. This is because the symptoms associated with tertiary syphilis occur as a result of the immune system of the host rather than as a direct result of the bacteria. A person develops severe complications as a result of a hyperimmune and inflammatory response against the pathogen. Because the symptoms don't develop as a result of active infections, antibiotics will not help eliminate the symptoms of this stage of disease.

Chapter 25 Applied and Environmental Microbiology

Multiple Choice

1. d	5. d	9. b	13. b	17. a
2. b	6. b	10. c	14. a	18. b
3. d	7. d	11. d	15. d	19. c
4. c	8. b	12. d	16. d	20. b

Fill in the Blanks

1. food, industrial, Environmental
2. starter cultures
3. intrinsic, extrinsic
4. water, allicin, microbial enzymes, garlic
5. food infection, food intoxication
6. potable, polluted
7. recalcitrant
8. bioremediation
9. populations, guilds, communities
10. competition, antagonism, cooperation, competition
11. nitrification, *Nitrosomonas*, *Nitrobacter*
12. sulfate, hydrogen sulfide
13. eutrophication
14. physically, chemically, biologically
15. smallpox, foot-and-mouth disease (FMD), fungi
16. surveillance, response protocols

Short-Answer Questions for Thought and Review

1. In ancient times, prior to water treatment, there were no guarantees as to the potability of water. Being fermented meant that wine and beer were safer to drink and could be stored without much spoilage.
2. The major difference between food fermentations and industrial fermentations is one of scale. Industrial fermentations are done on a much more massive scale than food fermentations, which can be done in homes as well as at food processing centers. Both need strict quality control, but in some cases there

will be more quality control associated with industrial fermentations, particularly with regard to pharmaceuticals. The fermentation products of food fermentations, acid in many cases, are left with the food to continue to restrict growth of harmful organisms. This is not so with pharmaceuticals, in which one particular compound is being produced and isolated. It is important that these compounds be pure so as not to accidentally make someone taking them sick.

3. Microbial fuel sources are so attractive because they offer a completely renewable, easily managed, and relatively low-tech alternative to oil. They can be grown on the wastes of other technologies, grown to high yield, and harvested relatively easily or set up in fermentation systems where their "fuels" can be collected during growth; and because they are producing natural substances, pollution risks would be lowered.
4. Large-scale environmental changes affect microbial communities by altering the conditions under which the community was established. Within the community, some populations will be better able than others to respond to the changes in the broader environment; these populations will survive while others may die out. Overall, changes to the environment lead to succession within the community as some populations come to dominate and others recede. Succession can be good, but it can also be bad, depending on whether or not the environmental change is natural.
5. Recycling helps keep the carbon cycle going by reducing carbon loss. Recycling and reusing recalcitrant materials limits the amount of carbon that is used to create more recalcitrant materials (reusing plastics means you don't have to keep making new plastics). By recycling, there is less loss of carbon and so more carbon remains in organic, degradable compounds that can be readily used by all living creatures.

Critical Thinking

1. Home canners are more likely to get food poisoning than people who buy industrially canned goods because it is much more likely that a home canner will make a mistake. If the food is not heated at a high enough temperature, for a long enough time, or the jars are not sealed correctly, then botulism can occur. As humans are more variable in their attention to detail, it is more likely that they will make mistakes than will the machines that control industrial canning processes.
2. In part, the fact that treated wastewater is not placed back into the drinking water supply stems from perception—most people consciously or unconsciously would not find the idea of drinking treated wastewater appealing, even if they know for a fact that it is clean. The second, more relevant, reason has to do with the fact that disease-causing organisms routinely find their way into wastewater, particularly as part of human wastes. Even if the treatment protocols are as efficient or even more efficient than those for drinking-water treatment, the small chance that something might slip through, particularly a virus, is enough to prevent the water from being reused for human consumption.
3. Natural bioremediation would be preferable because you would be using a naturally occurring organism that is present normally in the environment, rather than an engineered organism that has been created in a lab. The engineered organism might behave perfectly well in the environment, but there is a chance that, at some point in time, it could "escape." If this

happens, you have introduced an unknown microbe into the environment with no ability to predict what it will do. This is not the preferred method.

Concept Building Questions

1. If we have a microbe that is well studied, we can use the techniques of genetic engineering to modify it to do specific things. We can insert toxin genes, new genes ("good" copies of bad genes in humans), and maybe even, some day, drug genes or other similar genes that can give a microbe disease-fighting capabilities. By controlling the expression of ligands, we can control which cells the new microbe will attack and can direct it against tumor cells just as we could direct it against other microbes. Through the process of modifying a known platform, we could design better vaccines also. Essentially, the more we know about microbes and how they work in the body, the more we can manipulate them to work against the disease process.
2. It is more likely that new pathogens will arise from animal or human pathogens than soil or water microbes because of proximity and access and availability of growth requirements. Organisms in water and soil generally are free-living and are adapted to getting nutrition the hard way, surviving extremes in the environment and other factors they will not encounter in humans. They also will be very unlikely to have any sort of ligands with which to attach to human cells. Microbes in animals or humans, however, are adapted to living in other animals, will have ligands, will have nutritional features that allow them to exploit a host, and will be more likely to survive in these populations. Over time, mutations accumulate and so something new can arise; if it arises within the animal or human population it is already in, it is much more likely to remain there and start new diseases. Soil or water organisms would have to come up with too many new things to make a transition from free-living microbe to successful pathogen.
3. HIV could never be a bioweapon because of several factors. Its mode of transmission does not lend itself to delivery (there is no way to enable this fragile virus to survive outside of a human host); nor would it spread if released except by direct sexual contact, needles, etc. Its replication cycle does not lend itself to the "mass casualties" for which a bioweapon would be designed. Effects would not be seen for years and would not stand out relative to the HIV already present in the population (thus the terrorist would receive no recognition). Yes, there is fear associated with it, but not enough to overcome the two very real problems above. It can be diagnosed and to some extent treated, also limiting its usefulness as a weapon.